The Dutch implementation of the Data Retention Directive

310a

Onderzoek en beleid

The Dutch implementation of the Data Retention Directive

On the storage and use of telephone and internet traffic data for crime

investigation purposes

G. Odinot

D. de Jong

R.J. Bokhorst

C.J. de Poot

eleven international publishing

Wetenschappelijk Onderzoek- en
Documentatiecentrum
Ministerie van Veiligheid en Justitie

Onderzoek en beleid
The series *Onderzoek en beleid* comprises the reports of research commissioned and conducted by the WODC.
Inclusion in the series does not mean that the content of a report reflects the opinions of the Minister of Security and Justice.

Copies of this report and e-books can be ordered at www.elevenpub.com.

A limited number of free copies is available for civil servants of the Ministry of Security and Justice.
These can be ordered from:
WODC Library
P.O. Box 20301, 2500 EH The Hague
Free copies are available only for as long as stocks last.

The integral text of the WODC reports can be downloaded free of charge from www.wodc.nl.
Further information about other WODC publications is also available at www.wodc.nl.

Printed in The Netherlands

ISBN 978-94-6236-453-0
ISBN 978-94-6274-116-4 (e-book)
NUR 820

Foreword

In September 2009 the Data Retention Directive was adopted in the Netherlands, following the European guidelines on data retention. This law guarantees that telecommunications data of importance to investigation and prosecution of criminal offenses can be retained for a certain amount of time, thereby allowing it to be available for the investigation of serious crimes.

The European guideline was received well by all its member states. Though it seems evident that historical data about telephone and internet traffic can play an important role in criminal investigations, the privacy sensitivity of retaining these data poses a recurring point of debate.
The usability of telecommunications data for investigative purposes has increased dramatically due to the enormous rise in the use of mobile phones and smartphones. Telecommunications traffic often give a good picture of where people go and what they do, but this also means that their privacy is at a greater risk. Retaining these data therefore poses a greater threat to citizens nowadays than it did in the past. For this reason it is important to study how the data that must remain available according to the Data Retention Directive is retained, stored, secured and destroyed and how this process is monitored. In addition it is important to gain more insight into how these data are actually used in investigative practice. For example: when and by whom can the data be accessed and what role does that information play in investigation and prosecution of crimes?

This report provides a broad view of the way in which the Dutch Data Retention Directive is structured and how retained data is used in investigative practice. In order to do this, approximately forty professionals were interviewed in addition to accessing several other sources. On behalf of the authors, I would like to thank all those who contributed to this study: the interviewees; those who gave us access to data and those who provided us with information. We would also like to thank Nora Al Haider and Priya Soekhai, who scored the data, and Ruud Kouwenberg for his help in transcribing interviews. Finally we would like to thank the members of the Advisory Board who, with their critical questions and constructive remarks, provided valuable contributions to this study.

Prof. dr. F.L. Leeuw
Director, Research and Documentation Centre (WODC)

Table of contents

Abbreviations

AIVD	General Intelligence and Security Service (*Algemene Inlichtingen- en Veiligheidsdienst*)
AT	Telecom Agency (*Agentschap Telecom*)
BOB	Special Investigative Powers (*Bijzondere Opsporingsbevoegdheden*)
BoF	Bits of Freedom
BVH	Basic Enforcement Provision (*Basisvoorziening Handhaving*)
BVO	Basic Investigation Provision (*Basisvoorziening Opsporing*)
CBP	Data Protection Agency (*College Bescherming Persoonsgegevens*)
CBS	Central Bureau of Statistics (*Centraal Bureau voor de Statistiek*)
CIE	Criminal Intelligence Unit (*Criminele Inlichtingen Eenheid*)
CIOT	Telecommunications Research Information Center (*Centraal Informatiepunt Onderzoek Telecommunicatie*)
CvPG's	Board of Procurators General (*College van Procureurs-generaal*)
DCS	Digital Communication Traces (*Digitale Communicatie Sporen*)
EDPS	European Data Protection Supervisor
ECHR	European Court of Human Rights
FIOD	Fiscal Intelligence and Investigation Services (*Fiscale Inlichtingen- en Opsporingsdienst*)
IMEI	International Mobile Equipment Identity
IMSI	International Mobile Subscriber Identity
JBZ-raad	Home Affairs and Justice Council (*Raad Justitie en Binnenlandse Zaken*)
IP	Internet Protocol
ISP	Internet Service Provider
ITU	International Telecommunication Union
KLPD	National Police Services (*Korps Landelijke Politiediensten*)
MIVD	Military Intelligence and Security Services (*Militaire Inlichtingen- en Veiligheidsdienst*)
MvT	Explanatory Memorandum (*Memorie van Toelichting*)
NAT	Network Address Translation
NAW	Name, Address and Location (*Naam, Adres en Woonplaats*)
NFI	Dutch Forensic Institute (*Nederlands Forensisch Instituut*)
NMa	Dutch Competition Authority (*Nederlandse Mededingingsautoriteit*)
OM	Public Prosecutor Service (*Openbaar Ministerie*)
OPTA	Independent Postal and Telecommunications Authority (*Onafhankelijke Post en Telecommunicatie Autoriteit*)
OvJ	Public Prosecutor (*officier van justitie*)
RC	Investigative Judge (*rechter-commissaris*)
SIM	Subscriber Identity Module
Sv.	Code of Criminal Procedure (*Wetboek van Strafvordering*)
TGO	Large Scale Investigation Team (*Team Grootschalig Onderzoek*)
TNO	Dutch Organisation for Applied Scientific Research (*Nederlandse Organisatie voor Toegepast-natuurwetenschappelijk Onderzoek*)

Tw	Telecommunications Act (*Telecommunicatiewet*)
ULI	National Interception Unit (*Unit Landelijke Interceptie*)
VoIP	Voice over IP
WBP	Privacy Protection Act (*Wet Bescherming Persoonsgegevens*)
WOB	Public Nature of Government Act (*Wet Openbaarheid van Bestuur*)
WODC	Research and Documentation Center (*Wetenschappelijk Onderzoek- en Documentatiecentrum*)
zwacri	serious crimes (*zware criminaliteit*)

Summary

The study: background, research questions and data collection

Background to the research questions

The Dutch implementation of the Data Retention Directive was adopted on the 1st of September, 2009. The main reason for the storage of call detail records of telephone and internet traffic data is its potential in the aid of the investigation and prosecution of serious crimes. For example, this type of data can be used to ascertain the time and place at which a particular mobile telephone was used to make a call. The data also makes it possible to find out whether and when a computer or mobile telephone made an internet connection. Telecommunication traffic data can be used in cases involving a crime that merits pre-trial detention, a reasonable suspicion of a crime being planned or committed in an organized context and indications of a terrorist offence.
However the fact that this data has to be stored for a certain period of time is a recurring point of debate. There is a need both in the Netherlands and at European level (EU 18620/11) for a clearer understanding of how the police and judicial authorities use the data kept under the Telecommunications Data Retention Act (referred to below as 'the Act').

The purpose of this study is to clarify how the Act works in practice. This study extends beyond the scope of an evaluation process (cf. Wartna, 2005; Nelen et al., 2010), because there is a need not only for an understanding of how the Act has been shaped in practice but also of how the data to be kept available under this Act is actually used for criminal investigations in practice.
However, it is not possible – as it would be in a product or effect evaluation – to ascertain how the introduction of the Act has affected the use of traffic data in criminal investigations. The telecommunication data at issue here was already available for criminal investigation purposes before the Act was introduced, and was already being used in criminal investigations into serious crimes prior to the introduction of the Act.
Although the Act has resulted in the retention periods being harmonised, the fact that other changes have taken place in the meantime means that it is only barely possible to measure and identify any possible effects thereof. Changes in how telecommunication data is used in practice can be attributed primarily to the emergence of mobile and 'smart' telephones and to the increased accessibility of internet communication. It is thus easier to look into the use of telecommunications data in criminal investigations than it is to relate the findings to the introduction of the Telecommunications Data Retention Act.
This study focuses both on questions about how the Act has taken form and questions about how the retained data is used in practice.

In this context, there are various organisations and parties involved in the storage, maintenance, and use of telephone and internet traffic data. The providers are required to retain and secure the data, keep it available for investigative purposes and to destroy it at the prescribed time. This process is regulated by the Telecom Agency (*Agentschap Telecom*). The Dutch Data Protection Authority (*College Bescherming Persoonsgegevens*) has the more general task of regulating the use of privacy sensitive data. The Police and the Public Prosecution Service use this data for the investigation and prosecution of serious crimes, and the judiciary uses it in the legal decision-making process. The emphasis of this report lies on how the retained data is used in practice, thus providing a clearer understanding of the usefulness and necessity of the retention obligation. The complexity of how the Act works in practice is reflected in the descriptions of how the various parties perform their tasks. This report provides fairly detailed information about how the stored data is used in practice. Other parties are touched upon, but do not form the main focus of this study.

Data collection

Various methods have been used to answer the research questions. In addition to conducting an extensive review of the literature (national and international), both qualitative and quantitative data on the use of historical traffic data was collected from organisations such as the National Interception Unit (*Unit Landelijke Interceptie*) of the national police services, the Dutch National police, the judiciary (Public Prosecution Service) and the legal profession. A desk study was also carried out involving the examination of legal texts and their explanatory notes, secondary legislation, parliamentary papers, implementing agencies' written documents, and scientific literature. Also, 17 face-to-face interviews and 16 telephone interviews were conducted for the study, which involved speaking to a total of 41 people in the period from June to October 2012. Finally, court judgements were analyzed to ascertain how the Dutch courts used retention data as evidence in criminal trials.

Remote communication, developments and implications

In recent years the mobile telephone has been replaced by the smartphone, and many people are online 24/7 these days. The use of smartphones means that people are much more likely to communicate in the form of short messages via apps and email, and phone calls are being made increasingly online as well.
Technological innovations and the accompanying fragmentation of communication and the use of various online services makes it difficult to keep track of all of a person's remote communication. Additionally, not all traffic data

that is generated is covered by the law. Many internet users have email accounts with webmail services such as Hotmail, Gmail or Yahoo, which are provided by a foreign company. Consequently, the data is not necessarily retained for Dutch criminal investigation purposes. The same applies to providers of services in the *cloud*. In cases where investigative services want to obtain traffic data from foreign suppliers nonetheless, they need to submit a request for legal assistance and have to wait and see whether the data is still available.

The legislative history and European regulations on the Data Retention Directive

Partly in response to the terrorist attacks in Madrid in 2004 and in London in 2005, 3 May 2006 was the introduction date of the EU Directive aimed at guaranteeing that certain telecom and internet data are retained and kept available for the investigation and prosecution of serious crime.

Retained data

Section 5 of the Directive stipulates the categories of data to be retained with regard to aspects including the designation, the date, the time and the duration of the communication. It is not permitted to retain data from which the content of the communication can be derived. The Member States were required to convert the Directive into national legislation by 15 September 2007; an extension was given until 15 March 2009 for the obligation to retain internet data. Not all the Member States have converted the directives into legislation. The term 'serious crime' has not been defined in the directives. This is reflected in the various grounds laid down in the legislation of the Member States that facilitate access to the retained data for criminal investigation and prosecution purposes. As with the duration of the retention period, the harmonisation envisaged by the EU legislation has only been achieved to a limited extent.

Privacy

The Act affects the privacy of members of the public. In the first place, the storage of telecommunication data involves a risk of unauthorised persons – such as hackers – gaining access to that data. A second, different type of breach takes place as soon as the police and judicial authorities are granted access to retained data in the context of an investigation. According to the ECHR(2008, 30562/04) it is permissible to limit the right to privacy only if provided for by law and necessary in a democratic society.

The Dutch Code of Criminal Procedure (CCP) stipulates who has access to the retained telecom and internet data and under which conditions. The Public Prosecutor can claim the issue of traffic data (Article 126*n* and 126*u* CCP) if there is a suspicion of an offence that merits pre-trial detention or a reasonable suspicion that crimes are being planned or committed in an organised context. An investigating officer can claim identifying data (Article 126*na*, 126*ua* CCP). The details that can be obtained are what are known as the user details (name, address, place of residence, number and type of service). If there are indications of a terrorist offence, the Public Prosecutor can obtain traffic data (Article 126*zh* CCP) and an investigating officer can claim user data (Article 126*zi* CCP). For an exploratory investigation into terrorist offences the Public Prosecutor can also claim databases of public and private bodies in order to have their details processed (Article 126*hh* CCP).

The retention and securing of the data in practice

The regulatory authorities

Compliance with the rules is supervised by the Telecom Agency Netherlands, which operates as an independent regulatory authority and supervises compliance with the Act. The Telecom Agency is a division of the Ministry of Economic Affairs and reports directly to the Minister of Economic Affairs. Additionally, the Dutch Data Protection Authority regulates all statutory regulations concerning the retention, use and processing of personal data.

The providers

Meetings were held with four providers in order to gain an understanding of how they approach the obligations under the Act. Prior to the retention obligation being introduced the retention periods varied between companies. Despite the Act's long start-up period, its implementation proved to be a sizeable project for the large providers.
The two large providers interviewed for this study, maintain a database filled with data to be retained under the Act. This data is automatically destroyed when the retention period ends. A small provider interviewed for this study only recently actively started operating the retention periods because the quantity of data to be retained became too large. When they receive a request, the data applied for has to be taken manually out of the system by an employee.
The government has reached an agreement with the large Dutch suppliers concerning compensation for the personnel needed to issue data retained under the various Acts and government regulations. Small providers are not covered by this arrangement.

The owners of a fourth interviewed supplier recognise themselves in the documentation of the Telecom Agency as parties obliged to retain the traffic data of the email services they offer, but indicate that they do not comply with this for idealistic reasons. The researchers have asked the Telecom Agency whether the services offered by this company are subject to the retention obligation. According to the Telecom Agency they are not, but it acknowledges that certain parts of the legislation have become unclear owing to technological innovations.

Regulatory authority

The Telecom Agency also oversees the implementation of operational processes. The supervision is provided for in a monitoring cycle in which the data suppliers are questioned about how they retain, secure and destroy the data. However, the Telecom Agency does not have the instruments and powers to monitor the content of the retained and delivered data. Section 18.7 (2) of the Dutch Telecommunications Act expressly stipulates that the regulatory authority is not authorised to retrieve traffic or location data retained by the providers under Section 13.2*a* of the Telecommunications Act.

The use of historical traffic data in practice

The Act makes a clear distinction between telephony and internet traffic data. To be perfectly clear, this report maintains that distinction. But in practice the distinction has virtually faded away and experts feel that the Act operates an incorrect division into two categories.

What is retained?

The appendix to Section 13.2*a* of the Telecommunications Act contains a summary of the telephone data to be retained. This data includes the number of the caller and the party called, the time and duration of the call and the location. This data must be kept for a period of one year. The content of a call or an SMS is not subject to the retention obligation. The traffic data of the sent or received message is subject to that obligation. Attempted calls in which no connection is made fall under the retention obligation as well.

What is at stake?

According to crime investigation professionals historical traffic data is retrieved in virtually all larger criminal investigations in which suspects or victims may have used their telephone. In 2012 the number of claims for the disclosure of telecommunication data totalled to 56,825.

These claims were used to obtain information about the use of the telephone and possible IP-traffic, such as: the number that was used to make the call, when the call was made, the duration of the call and from which location, and whether there was any online contact. This information plays an important and highly valued role in criminal investigations. If an investigating team wants to obtain traffic data, it has to obtain the approval of the Public Prosecutor. The investigating team has to indicate what it is seeking to achieve with the information, and obtaining the information must be proportional and observe the principle of subsidiarity. The intentions of the investigating teams in obtaining traffic data can be placed in a number of categories: (1) to identify a user, (2) to establish contacts, (3) to determine a location, (4) to trace an IMEI number, and (5) to make a decision on capacity before wire tapping.

Relevance and retention period of telephony data

All of the interviewed professionals and experts said that they found historical data on telephone traffic to be highly relevant. A number of interviewed crime investigation professionals indicated that they not only wanted to obtain the start location (*first cell*) of a telephone call, but also the end location (*last cell*). However, the location where a call ends, i.e. the final connection with a transmission tower, is not stated in the appendix to Section 13.2*a* of the Telecommunications Act.
It emerged from the interviews that most of the professionals and experts among the police felt that the one-year retention period is sufficient for the work that they do.

Historical internet traffic data

What is retained?

Historical traffic data concerning internet and email usage can yield information about matters such as the IP addresses someone has used, and the email contacts of the sender and receiver. The content of calls, messages or emails and search terms entered in a search engine and the IP addresses of searched internet pages are not covered by the retention obligation.

Relatively little deployment

During the interviews conducted for this study, it became clear that the criminal investigation professionals had little or no knowledge of how historical data concerning internet traffic could be used for crime investigation purposes. Additionally, the work related to internet matters is often carried out by

experts because the digitisation of today's society does not yet form part of the day-to-day work of many investigating officers. At the same time we established that technological developments move at a very fast pace. So fast that it is difficult even for the scarce experts to keep up with them.
Historical internet traffic data is often retrieved in response to a crime or offence committed with the aid of or via the Internet, such as sending threatening emails, internet fraud, human trafficking and the distribution of images of child sex abuse. The most important reason given for retrieving data is to *identify a user* or a connection. Fixed IP addresses usually remain unchanged for longer periods and the use can easily be traced either at the provider or at the Telecommunications Research Information Centre (*Centraal Informatiepunt Onderzoek Telecommuniatie*). However, identifying a mobile internet user on the basis of historical traffic data is a laborious process and in many cases not possible.

The relevance and retention period of internet data

According to various experts the majority of data described in the appendix to Section 13.2*a* of the Telecommunications Act is out-dated. The regulation is no longer in keeping with today's internet usage or with the technological developments that have taken place in this area since the Telecommunications Act was introduced in 2009. This has led to the retention of data of members of the public that is not or is only barely used by the criminal investigation services. A meticulous review of the regulation governing IP traffic and the retention of IP data therefore appears appropriate.
The professionals and experts interviewed for this study, who are familiar with the internet traffic data, all believe that the six-month retention period is too short; there is clearly a need for IP traffic data that goes back further in time for criminal investigations into offences for which this data is retrieved.

The retrieval of transmission tower data

Retrieving traffic data based on a location yields information about all mobile telephones which, in the indicated time frame, have been called, have made calls or had an internet connection via the tower location in question. For permission to retrieve transmission tower data there must be a suspicion of an offence as specified in Article 67 (1) of the Code of Criminal Procedure and the use of the data must be in the interest of the investigation.
Transmission tower data is retrieved mainly for serial offences. In such cases the data of various locations are compared, with the intention to pinpoint a recurring number. Of course, this investigation method only has a chance of success if the suspect used his telephone around the time of the offence.

Alternative?

Opponents of the retention obligation regard the targeted freezing of data as being a less privacy-violating solution because this involves a specific data set that is retained for longer, rather than retaining all the data of all of a provider's customers. None of the experts we spoke felt that freezing data was a comparable or equivalent alternative to a general retention obligation because this would rule out the possibility of retrieving data retained a longer time ago. To be able to use this data it is necessary to know in advance – while the data is still available and can be frozen – what data will be needed at a later date. Given that it is sometimes not until later that offences come to the knowledge of the police, and suspects are sometimes not identified until long after a crime has been committed, it is necessary to retain this data for later use in the criminal investigation process.

The use of traffic data in figures

The Telecommunications Act makes it compulsory to annually publish the number of data requests about telecommunications traffic data made by criminal investigation services (Section 13.4 (4) of the Telecommunications Act). In 2012 a total of 56,825 claims for the disclosure of traffic data were made. However the number of claims announced by the Minister also includes data not covered by the Telecommunications Data (Retention Obligation) Act.
It should also be noted that the retrieval of telecom data in the Netherlands is registered by telephone number, IMEI number, IP address or 'transmission tower location' on which data is retrieved. These figures do not provide an insight into the number of people whose telecommunication data is retrieved each year, or the number of criminal investigations or the nature of the investigations for which the data was retrieved. Neither do the figures provide any insight into the extent to which a claim has actually resulted in data being issued.

Court rulings

This report also provides an insight into the use and value of traffic data in court rulings. A total of 74 rulings dating between July 2012 and February 2013 were found in which the term historical traffic data concerning telephony occurred. This data was generally used in the rulings to demonstrate 'contact between suspects' and 'locations'.
A search of court rulings in which IP traffic data was used in the judgement, yielded 26 rulings in the period from January 2009 to February 2013. This IP data was mentioned mainly in the rulings concerning criminal

investigations into child pornography. More than half of the judgements concerned the downloading and/or distribution of images of child sex abuse. The retrieval of this data is not so much about where the suspect was and with whom he communicated, but rather whether the suspect could be linked to the internet address that was used or other user data.

1 The Dutch Data Retention Directive – an introduction

On 3 May 2006 a European directive to retain certain telecommunications data came into force, ensuring that such data are available for police and judicial investigations into serious crimes. The Telecommunications Data Retention Directive applies to companies that provide internet services as well as telephone companies (hereafter referred to as providers). Though the police and prosecutors service have long had the right to exact this type of information from providers, those same providers were obligated by law to destroy any user information as soon as it was no longer needed for its own business operations. As a result, information needed for investigative purposes sometimes was no longer available.[1] The Data Retention Directive includes provisions for the storage and securing of telecommunications data, the supervision thereof, and the legal protection of those whose data are stored as well as the data itself.

The central idea behind the requirement to retain telephone and computer data is their use in investigative purposes. On the basis of that traffic information it is possible, for example, to determine when and where a particular (mobile) phone has been used. It is also possible to determine whether and when a computer or mobile phone has connected with the Internet. In the case of a smartphone with mobile internet it is even possible to trace the location from where the data traffic took place. Thus, by requesting retained telecommunications data, it is often possible to track the movements of suspects under investigation retrospectively. According to the data retention directive providers are required to store the telecommunications data for a period ranging from a minimum of six months to a maximum of two years. Member states are free to choose the exact duration of mandatory retention within this range.

Not all EU member states welcomed the Data Retention Directive. Questions were also raised in the Netherlands about the usefulness and necessity of this retention directive. For example, does the Data Retention Directive contribute significantly (or even sufficiently) to the investigative process? In addition there has been criticism concerning the (dis)proportionality of the act in terms of infringement on civil liberties. There has also been controversy concerning possible conflicts of the Data Retention Directive with Article 8 of the European Convention on Human Rights and Article 7 of the Charter of Fundamental Rights of the European Union.[2] As a result, some member states, such as Sweden and Austria, have not ratified the directive. In the Czech

1 There was a limited obligation for providers to retain data. Traffic data of customers whose names and addresses were not known (e.g. prepaid numbers that can be used anonymously) were obliged to be kept for three months.

2 The European Data Protection Supervisor called the European directive 'the most privacy invasive instrument ever adopted by the European Union' (Hustinx, 2010). The Electronic Frontier Foundation regards the directive as 'the most prominent example of a mandatory data retention framework', (www.eff.org, accessed on December 7, 2012), and Frost and Sullivan (2010, p. 15) described the directive as 'the most extensive piece of data retention legislation adopted by any country or union of countries today' (see also Hathaway & Klimburg, 2012, p. 40).

Republic, Germany and Romania the constitutional courts have even annulled legislation aimed at ratification of the directive.[3] The mandatory retention period in countries that have ratified the Data Retention Directive varies greatly within the directed norm (six months to two years).

Dutch law concerning ratification of the retention directive
Legislation pertaining to the ratification of the Data Retention Directive is in effect in the Netherlands, and the Telecommunications Data Retention Act, here referred to as the retention act, entered into effect on September 1, 2009. As mentioned above, a legal basis for the retrieval of telephone and internet traffic data already existed in the Netherlands.[4] The retention law hasn't changed this. Both before the ratification as well as thereafter a disclosure claim can (and previously could) be made by investigative authorities in cases when pre-trial detention is permitted, where reasonable suspicion exists that crimes are planned or committed in an organized context, and when there are indications of a terrorist offense.[5] What has changed, however, is the retention period. Because telecommunications data is required to be stored longer post-ratification, it is now possible to obtain information covering a longer period of time. This is a direct result of the new legislation. In addition, more traffic information will be available post-ratification because providers are now legally obliged to store these data specifically for these purposes. Moreover, the ratification of the Retention Act brought with it requirements pertaining to the supervision of the stored data, including the security of stored data as well as its destruction. Pre-ratification providers could store data unsecured and indefinitely.
In the draft bill, it was suggested that communications data pertaining to both fixed and mobile telephony, as well as data pertaining to internet should be stored for eighteen months. The House of Representatives reduced the retention period to twelve months. Thus, historical communications data is now only available for disclosure for a maximum of twelve months.
The twelve month retention period for internet data led to much debate in the Netherlands. Hearings on this subject have been held with experts from the field and an extensive exchange took place between the then Minister of Justice and the Senate. In that context, foreseeing that telephoning via the Internet would increasingly replace 'traditional' telephoning, the Minister argued that the retention period for internet data should be twelve months in

3 Evaluation of the European Commission on European Data Retention Directive (COM (2011) 225 final).
4 A distinction is made in the law, between information pertaining to user communications (or traffic) (Article126*n/u/zh* of the Dutch Code of Criminal Procedure, CCP) and information pertaining to the user as in name, address, postal code, city, number and type of user service (Article 126*na/ua/zi* CCP). These two categories of data are referred to as traffic data and user data respectively. In general, the authority to exact traffic data includes the authority to exact user data.
5 The Dutch Code of Criminal Procedure distinguishes between three types of crime investigation; Article 126*n* concerns the retrieval of traffic data for the purpose of investigating crimes committed; Article 126*u* concerns the retrieval of traffic data for the detection of crimes planned or committed in an organized context; and Article 126*zh* concerns the retrieval of traffic data for investigations based on indications of a terrorist offense. User data can be claimed on the basis of Article 126*na*, 126*ua* and 126*zi* CCP respectively.

light of technological developments.[6] However, due to the opposition, this position proved untenable for the Senate, despite much deliberation. As a result, the Minister agreed to amend legislation, reducing the retention period for internet data to six months following the ratification of the retention act by the Senate. The legislation was amended to that effect[7] in July of 2011.

Providers of communication services can only be required to disclose traffic data if there indeed is or has been communication traffic. In other words, a user must actually make a connection, or an attempt thereto, between his telephone or other telecommunications device and that of another. In theory it is possible for telephones in standby mode for example, to generate cursory data pertaining to location. This information is however, exempt from the disclosure directive under Article 126*n*/ *u*/ *zh* of the Dutch Code of Criminal Procedure (CCP), because no active connection, or attempt thereto, is made.[8]

The storage of telecommunications data

Data resulting from the use of a telecommunications service consist merely of content data, traffic data, location data and other information, such as subscriber identification data (name and address). Data on the content of the communication may not be stored. On the basis of the Data Retention Act it is thus not permitted to store information pertaining to a user's internet surf history or view the contents of user emails. The historical telecommunications data that can be stored pertains to the time at which someone used a particular telephone (number) to connect with another number, the duration of that connection, and the location from where that conversation was made. Name and address of both caller and called are also subject to disclosure. With regard to the internet, historical data can be obtained pertaining to the time and duration of connection between a computer or mobile telephone and the Internet, as well as information about the location from where the connection was made.

Traffic data stored under the Data Retention Act are stored decentralized, by the individual providers, who are in turn responsible for the storage, security and destruction of information. The Telecom Agency (*Agentschap Telecom, AT*), operating in a supervisory capacity, oversees that all rules concerning data retention are abided by. To this effect, the Telecom Agency uses the Independent Postal and Telecommunications Authority's (*Onafhankelijke Post en Telecommunicatie Autoriteit, OPTA*)[9] database working together with

6 *Preparatory Memorandum (Senate)*, 2008/09, 31 145, C. p. 8.

7 *Stb.* 2011, 350.

8 Cursory data on the location for a telephone in standby mode may be claimed on the basis of Articles 126*ng/ug* jo., 126*nd/ud* and 126*ne/ue* CCP.

9 The Netherlands Consumer Authority (*Consumentenautoriteit*), the Netherlands Competition Authority (*Nederlandse Mededingingsautoriteit, NMa*) and the Netherlands Independent Post and Telecommunications Authority (OPTA) joined forces on April 1 2013, creating a new regulator: the Netherlands Authority for Consumers and Markets (*Autoriteit Consument en Markt, ACM*).

the Dutch Data Protection Authority (*College Bescherming Persoongegevens, CBP*). The supervision is organized in an annual cycle of risk identification and appropriate action. In addition, the Telecom Agency strives to visit all approximately 600 registered internet and telecom providers in the Netherlands at least once every four years. Requests for the disclosure of the stored data can be made by investigative agencies for investigative purposes through the National Interception Unit (*Unit Landelijke Interceptie, ULI*) of the National Unit (*Landelijke Eenheid*).[10] The stored data can be made available for up to one year prior to the claim for disclosure for telephone data and up to six months prior for internet data.

The number of claims in the Netherlands

The Minister of Security and Justice published the number of claims for the disclosure of telecommunications data for the first time in 2010. In the second half of 2010 24,012 claims were submitted and in 2011 that number increased to 49,695. In 2012 the total number of claims submitted to 56,825. Generally, the information requested pertained the use of telephone and Internet Protocol (IP) numbers such as: with which number was called? When were calls made? How long did telephone calls last and what was the location of the caller at that time? Did the user connect with the Internet? Investigation services use this historical traffic data for different purposes. For example, in murder cases with an unknown perpetrator usually the historic telephone traffic of the victim is requested, in order to get an idea of who (or at least with which phone) the victim was in contact with in the period prior to his or her death. In the case of criminal collaboration, the historical traffic data can provide information about whether or not suspects had contact, when such contact was made, and from where conversations were held. Statements made by suspects pertaining to contact with potential co-suspects can also be checked using telecommunications traffic data. In addition, historical telecommunications traffic data can be used to better understand social networks by providing insights into the location from where certain persons have had contact with each other. Moreover, this type of traffic data can show digital traces left by perpetrators, thereby sometimes leading investigators to (new) suspects. For example, by retrieving information from the transmission tower nearest to one crime scene (e.g. location of a robbery), and comparing it to information retrieved from a transmission tower nearest to another crime scene (e.g. a burnt out getaway car), it is possible to determine a possible suspect when the same numbers are found at times corresponding to the crimes at both transmission towers near to the two crime scenes. Precursor to this is obviously that the perpetrator has actually used his telephone in the area of each crime scene. Finally, the use of historical traffic data can play a role in deciding whether or not to use more

10 At the time of this research, the National Interception Unit resided under the National Police Services (*Korps Landelijke Politiediensten, KLPD*). As of January 1, 2013 this unit has become part of the National Unit.

invasive methods of investigation, such as tapping a particular phone number.

Privacy breach versus investigation interest
It is evident that historical telecommunications data can be useful for the investigation and prosecution of criminal offenses. However, the fact that this type of data must be retained is a recurring point of discussion. There is a need for more insight into the use of data stored under the Telecommunication Data Retention Directive, both in the Netherlands as well as at level of the European Union (EU 18620/11). In response to questions posed by the Senate pertaining to the usefulness and necessity of the Data Retention Act, concerns as to the protection and security of stored data, and regarding the harmonization of data retention directives at European level, the Dutch Minister of Security and Justice agreed to commissioning a study of the Dutch Data Retention Directive, addressing, amongst others, these questions.
On April 18, 2011 a report issued by the Commission on European Data Retention Directive (COM (2011) 225 final) concerning the European Data Retention Directive was published. Its main conclusion was that the Directive is a valuable investigative tool, which should be upheld. It is the opinion of the European Commission however, that there is too much variation between the member states in the purpose of retaining data, retention periods and type of data stored. For this reason, the European Commission is looking into the possibilities to harmonise these issues between EU member states. Both the Senate as well as the House of representatives in the Netherlands have expressed criticism of the European Commission's evaluation report (E110022), suggesting that the usefulness and necessity of the Data Retention Directive were insufficiently addressed. Moreover, the directive offers too little insight into new forms of communication that fall outside the scope of the Data Retention Directive. Both the Dutch Senate and the House of representatives criticise that too many questions concerning the Directive in practice remain unanswered. Several political parties even campaigned to have the Directive repealed (E110022).
The European Data Protection Supervisor (EDPS) also criticized the European Commission's evaluation report. In an article published on May 31, 2011, the EDPS suggested that the Data Retention Directive does not meet the fundamental rights to privacy and data because: (1) the need for data retention as described in the Directive is insufficiently established, (2) data retention could be arranged in ways that pose less of an infringement on privacy, and (3) the Directive allows too much leeway for member states to determine for which purpose they want to use stored data, who can have access to the stored data and the circumstances under which access to the stored data is granted.
The aim of the present study is to gain insight into how the Data Retention Act works in practice. It will examine what data is stored in practice, how it is

stored, protected, secured and destroyed, and by whom and under which conditions the data can be retrieved. In addition, this report focuses on how the data stored under the Data Detention Act is made available for investigative purposes. Further, it provides insight into the utility of the retention periods and whether these periods are sufficient. Finally, the availability of other potential investigative procedures is examined, as well as whether or not data retention can be achieved in a less privacy intrusive manner.

This study thus focuses on the implementation of the Data Retention Act and how data stored under this Act is in fact used. Strictly speaking, this study is not an evaluation of the Data Retention Act, but rather goes beyond a process evaluation (cf. Wartna, 2005; Nelen et al., 2010). Not only is there a need to understand how the data retention law has been implemented in practice, but there is also a need to understand how these data are in fact used. How these data are used in the process of investigation, and the degree to which the legally determined retention periods suffice, lie at the core of this study. It is impossible to measure the effect of the implementation of the Data Retention Act as one might do with e.g. a product or impact evaluation, because the type of telecommunications data directed by this Act were in fact available for investigative purposes before the implementation of the Data Retention Act. Moreover, the implementation of the Retention Act may have resulted in a change in the use of these data for investigative purposes, but changes can also be attributed to the general increase in mobile telecommunication and internet (as a means for communication) use. It is therefore possible to examine how telecommunications data are used in the criminal investigation practice, but not to relate to the implementation of the Retention Act to those findings. Although the introduction of the data retention law ensured harmonization of retention periods, the effects of its implementation are hardly measurable due to other changes that have occurred in the meantime. We therefore chose for a broad research design including questions on how the law is designed as well as questions about the use of the stored data.
Several parties and stakeholders are involved in the storage, availability and use of data for investigative and prosecution purposes. The providers store, secure, make available and destroy data timely. The Telecom Agency oversees this process. The Data Protection Agency has the general task of monitoring the use of privacy-sensitive data. The Police and the Public Prosecutor use these data for the investigation and prosecution of serious crimes and the Courts use the information in the judicial deliberation. The Data Retention Law is de facto a complicated structure, which we have chosen to represent by describing how the different parties perform their tasks. Though other parties are discussed, they do not form the focus of this study. This report pays a relatively large amount of attention to the way in which the stored data is used.

1.1 Purpose and research questions

The aim of this study is to gain insight into how the Dutch Data Retention Directive is shaped, the way in which the data is stored in practice, and how such data is used by the police and the judiciary. We will focus on the following questions:

- How has the telephone and internet market evolved in recent years and what has the effect thereof been on the way in which historical telecommunications data can be used for investigative purposes?
- What is the purpose, the background and the content of the Data Retention Act in the Netherlands?
- Are the data that should be retained for investigative purposes under the Data Retention Act indeed stored?
- Who can access these data and under which circumstances can the data be retrieved?
- In which ways these data are protected against unlawful use and is the government compensation therefore sufficient?
- In which way the data stored are under the Data Retention Act used in criminal investigation?
- Which considerations and goals underlie the retrieval of historical telecommunications traffic data and what results can be achieved therewith?
- Are the statutory retention periods of one year for telecommunications data and six months for internet data useful, necessary and desirable or would the investigation process be better served by shorter or longer retention periods?
- Have other European member states implemented the Data Retention Directive?
- Are there any investigative procedures available, with which the same results can be achieved as with the retrieval of stored telecommunications traffic data, that pose less of an infringement of privacy for large groups of citizens?

1.2 Research design

This study has sought answers to research questions using different research methods. A review of the literature, as well as interviews were used to describe the developments in the telephone and internet market and its implications for the storage of telecommunications traffic data and their use for criminal investigations. A literature review of legal texts and explanatory notes, subordinate legislation, parliamentary papers, written documents supplied by implementing agencies and scientific articles was conducted in order to describe the laws and regulations concerning data retention. Further, Telecom Agency officials as well as several providers were interviewed

to gain insight as to how telecom and internet providers meet the legal requirements concerning the storage of data and protection against abuse. Existing literature on this subject was also studied.
In order to determine the usefulness and necessity of the Data Retention Directive, we looked at how the data stored under this directive were used in practice for investigation and prosecution. To do this several methods were used. Firstly, the professional literature, both national and international, on this topic was studied. Additionally both quantitative and qualitative data were collected concerning the use of historical traffic data. For this, data was collected from the National Interception Unit, the police corps, the judiciary, the Public Prosecutor Service and from the legal profession (solicitors). These data were collected regionally as well as nationally. Finally, court rulings were analyzed to see how data stored under the Data Retention Act were used as evidence in criminal cases. To this regard, our research provides a supplement to Mevis et al. (2005) investigative-dossier study on the usefulness and necessity of telecommunication data retention.

1.2.1 Interviewees

Many players are involved in the execution and enforcement of the Data Retention Act and the retained data can be used for different purposes. In order to provide a broad view as to how the Data Retention Act actually works, how the stored data is used and of the considerations underlying the retrieval of these data, interviews were conducted with experts and professionals in the field. Those interviewed for this study are on the one hand experts with specific knowledge of certain aspects of the storage, protection and retrieval of historical telecommunications data, or of the control and supervision of this process. On the other hand, persons interviewed came from a population of professionals in the field with a wealth of practical experience in the field of criminal investigation, investigation and prosecution of certain types of crimes and how historical telecommunications data is used in those cases. We spoke with experts from amongst others the National Interception Unit, The Dutch Forensic Institute, the Telecom Agency, and the Data Protection Agency. In addition, we interviewed two large and two small providers of telecommunications and internet services, as well as professionals employed by the police (primarily team leaders and analysts), the judiciary (Public Prosecutors Office), special investigation services, and lawyers (with a wealth of experience in a wide range of criminal cases). Finally, we interviewed some experts not connected to the above mentioned organisations.
In total 41 people were interviewed of whom 25 people face-to-face and 16 by telephone.[11] The following is an overview how many people were interviewed per organization:

11 17 face-to-face interviews were conducted, 8 of which with more than one key person present; 16 interviews were conducted by telephone.

- National Interception Unit (1);
- police (15);
- Fiscal Intelligence and Investigation Services (*Fiscale Inlichtingen- en Opsporingsdienst, FIOD*) (2)
- Public Prosecutors Office (2);
- The Dutch Forensic Institute (1);
- Lawyers (3);
- Providers of telecommunications and internet services (6);
- Telecom Agency (2);
- Bits of Freedom (BoF) (1);
- The Data Protection Agency (3);
- Scientists/legal professionals/specialists (5).

1.2.2 Method of the empirical study

For the face-to-face interviews we used a semi-structured questionnaire containing a number of fixed topics. This questionnaire was adapted to the job or position of the interviewee. In addition to the questionnaire, several themes warranted a more extensive discussion for most of the interviewees. The interviews took an average of one and a half hours. The telephone interviews were conducted using a structured questionnaire. This questionnaire was also adapted to the function and position of the interviewee and lasted on average half an hour. All interviews were recorded (audio) with the consent of the interviewee and subsequently transcribed. The transcriptions were anonymised and coded by two researchers. The code list included a detailed description of all items discussed during the interviews. Statements made by interviewees that were difficult to code were first placed in the 'other' category, and analyzed at a later date. Where relevant they were included in the report. This method allowed us to use MaxQDa, a software program for the analysis of qualitative data, to analyze statements of interviewees per specific topic or theme. The code list forms the basis for chapters 4 and 5 of this report. The quotations that appear in those chapters are not isolated but rather carefully selected to represent the experience and/or views of more interviewees where more persons were interviewed on the same topic. This does not apply to the quoted statements of experts, because they are sometimes the only ones who have commented on a particular topic.
The quotes serve to illustrate the topics described in the text.

1.2.3 Structure of the report

This report includes seven chapters in which first the changing world of telecommunications is described (Chapter 2). In Chapter 3, legislative provisions and conditions regarding the use of traffic and location data are described, and the question as to whether or not other European countries have implemented the Data Retention Directive is addressed. The retention and security of traffic and location data is described in Chapter 4. The experiences of providers of telecommunications and internet services is also expressed in this chapter and the monitoring of the implementation of the Data Retention Act is also described. Chapter 5 continues with a description of the de facto use and application for investigative services of the retained data. In addition, it covers traffic and location data of telephony and internet. Subsequently, Chapter 6 shows the figures concerning the use of historical traffic data. A survey of the amount of claims for disclosure of telecommunications traffic data is presented as well as court rulings where traffic and location data played a role in judgement. Finally, we offer a conclusion in Chapter 7.

2 Remote communication, developments and implications

The authority to request historical traffic data from telecom providers for investigative purposes was introduced in the Netherlands in 1926 (see also Koops, 2002). In 2000, this authority was adapted and transferred to the Special Investigative Authority Act (*Wet bijzondere opsporingsbevoegdheden*), a law that serves as the basis for the use of covert investigative procedures which infringe privacy. With the rise of the Internet, the authority that applied for retrieving data on telephone traffic also was used to request internet traffic data.
With the rise of the mobile phone and the widespread use of the Internet in the past decade, there was a large increase in the amount of traffic data stored as well as an increase in its utility. With the advent of the mobile phone the number of actual telephone calls made also increased significantly. Moreover, it has become possible to approximate where a mobile phone was located at the time a call is made.[12] By linking contact and location information both the privacy sensitivity of the historical traffic increases, as well as the usefulness of this information for crime detection and investigation. The development of mobile telephony also entailed that the privacy-sensitive information increasingly came into private company hands. This made it impossible for investigative service to recover data in all cases (see also Mevis et al., 2005). Data that were not recorded for their own operations could not be retrieved. Private companies were obliged to destroy all historical traffic data once they were no longer needed for their business operations.[13] The Data Retention Act changed this by requiring private telecommunications and internet providers to retain communication traffic data for a specified period of time for the purpose of investigation.
Because the Internet has become accessible to an increasing number of people, it has become an increasingly important communication tool. The mobile phone has been replaced in recent years by the smartphone, and many people are now online 24 hours a day, 7 days a week. The use of smartphones has led to the increase in communication through short messages via apps and email. People also call more and more via the Internet. This results in an increased amount of privacy sensitive information traces. At the same time, due to technological advances, an increasing number of traditionally available traffic data no longer falls under the Retention Act or is no longer readily available to investigative services. In the section below we will briefly outline the current telephone and internet market in order to provide a clearer picture of the context in which the Data Retention Action is to be regarded.

12 In order to make telephonic contact, a mobile phone must make contact with a transmission tower. Using the information concerning the frequency used, it is possible to determine from several kilometers up to several hundred meters accuracy (depending on population density), the location of that telephone at the time of calling.

13 There was a limited retention duty (see footnote 1).

2.1 The telephone market

The telecom market has changed dramatically in recent years. Mobile telephony has become an essential part of everyday life in the past decade. In the second quarter of 2012 there were 21.7 million mobile phone connections in the Netherlands in use and 7.1 million fixed telephone connections (Independent Postal and Telecommunication Authority, 2012). Previously the telecom market focused mainly on verbal communication, however increasingly this sector is focussing on data. Computers are no longer the only way to send data; mobile phones have taken on this function as well. In the second quarter of 2012, the Independent Postal and Telecommunications Authority[14] reported an increase in turnover of 12% from data usage and an increase in data volume by 21% compared to the second half of 2011 (Independent Postal and Telecommunication Authority, 2012). At the same time, the number of minutes telephoned has decreased steadily. In 2009, the number of minutes called per person from landlines decreased with more than 7%. Though mobile phone minutes called did increase in that year, according to the Independent Postal and Telecommunications Authority, the revenue thereof was not enough to offset the decline in fixed telephony (Independent Postal and Telecommunication Authority, 2009). In the first quarter of 2012 the total number of minutes called decreased from 5.8 to 5.7 billion (Independent Postal and Telecommunication Authority, 2012).
A mobile phone is no longer merely used to make calls, but also to send emails, to communicate via social media, take photographs, listen to music, play games, and for shopping and banking. For many of these activities an internet connection is required. The number of broadband connections in the Netherlands has surpassed 10 million and the vast majority of these connections (8.9 million) is a smartphone connection (Independent Postal and Telecommunication Authority, 2012).
The smartphone has become the most frequently used device for outdoor contact with the Internet. Figures from the Central Bureau of Statistics (*Centraal Bureau voor de Statistiek, CBS*) show that in 2012, more than 56% of all Dutch people aged between 16 and 56 used a smartphone to access mobile internet. Among young people aged between 12 and 25 years, 70% use a smartphone to go online outdoors almost daily.
In addition to the smartphone, the laptop computer is also often used for outdoor access to the Internet. The Central Bureau of Statistics explains that, due to the increased availability of Wi-Fi, laptop use has become more popu-

14 Since April 1, 2013 the Independent Postal and Telecommunications Authority has pooled forces together with the Consumer Authority (*Consumentenautoriteit*) and the Dutch Competition Authority (*Nederlandse Mededingingsautoriteit, NMa*) to form the new supervisory organization Consumer and Market Authority (*Autoriteit Consument en Markt, ACM*).

lar in 2012 in comparison to previous years. Particularly noteworthy is that the use of mobile internet in the Netherlands is well above the EU average.[15]

The Netherlands has a high quality infrastructure, which enables call and data traffic facilitation. There are six large telecommunications providers who own the network in the Netherlands. In addition, there are service providers and mobile virtual network operators that do not have their own network, but use the network of the major providers.
In order to be able to use fixed telephone lines, users must subscribe to one of the providers of the desired services. Subscription entails registration in order to ensure monthly payments. The traffic information generated by these users is easily coupled with an address. Similarly, for the use of mobile phones it is also possible to subscribe to one of the providers. In order to do this, the user will also have to identify and register himself with the provider. The provider subsequently supplies a SIM (Subscriber Identity Module) card which, once placed in the mobile telephone device, allows the user to make mobile phone calls. The provider thus charges the monthly rates accordingly. However, mobile phones can also be used without a subscription, and thus without the identity of the user being known by the provider. In the Netherlands 31% of the mobile phones used operate using a so-called prepaid SIM card (Independent Postal and Telecommunication Authority, 2011). This is a SIM card that represents a certain value. Once it is placed in the telephone, it enables connection to the provider's network. Prepaid SIM cards supplied by various telecom companies are sold in many places and credit can always be bought at a later date. This does not require registration, so user identity is not associated to the prepaid SIM card, and thus remains unknown to the provider. Traffic data from prepaid calls are thus anonymous and users of these cards are thus difficult to identify for investigators. When the calling credit has been completely utilised, the user can choose to buy new calling credit on the same SIM card. In that case, he retains his number and he remains customer with the same provider. For some users, however, maintaining the same number is of little importance. A phone number is not essential when the phone is mainly used to make calls and communicate via the Internet. By buying a new prepaid SIM card it is possible to re-access the network of a provider, whilst changing the phone number of the user. In short, by using prepaid SIM cards, it is possible to make calls and surf the Internet anonymously. Apart from the telephone number associated with the SIM card, each mobile telephone device has a unique International Mobile Equipment Identity (IMEI) number. Both SIM and IMEI numbers are stored and historical communications data of the two numbers can be requested by investigation services.

15 www.cbs.nl/nl-NL/menu/themas/vrije-tijd-cultuur/publicaties/artikelen/archief/2013/2013-3851-wm.htm (consulted on July 8, 2013).

2.2 The Internet

The Internet is a global computer network, of which the use has expanded enormously since the nineteen nineties. According to the Central Bureau for Statistics, in 2010 approximately 94% of Dutch people in the age from 12 to 75 had access to the Internet. Its use is popular. Two thirds of users view their email daily, almost 20% chat online, visit online communities daily and view online videos (ITU, 2011). Between 2005 and 2010 the number of internet users worldwide doubled. In a report by the Dutch Organisation for Applied Scientific Research (*Nederlandse Organisatie voor Toegepast-natuurwetenschappelijk Onderzoek, TNO*), it was estimated that in 2010 there were more than two billion internet users worldwide (TNO, 2010), suggesting that about 35% of the world's population was online. It is expected that in 2020 about 60% of the world population will be online (Hathaway & Klimburg, 2012).
The use of mobile internet is rising dramatically. According to the Independent Postal and Telecommunications Authority, the total data usage in the first six months of 2010 increased eightfold compared to the first half of 2008 (Independent Postal and Telecommunication Authority, 2011; TNO 2011). Market figures for the second quarter of 2012 show that this trend continues. The use of mobile internet is increasingly facilitated by the 'hotspots' that are available in more and more places. These are Wi-Fi networks to which a user as a customer can login to connect to the Internet, often free of charge. Because such hotspots are not regarded as public domain, they do not fall under the Data Retention Act. Wi-Fi networks are often found in the train, at stations, in restaurants, hotels and shopping malls, whereby customers can use the Internet upon logging in.
The Internet is used for shopping, looking up information, watching films and clips, making calls, sending emails, playing games, accessing social media and even for storing data. Storing personal data, as well as editing it, can be done in what is called a 'cloud'. This means that for example, photographs or other data are stored on servers and can actually be stored anywhere in the world, rather than on the local hard disk of a personal computer (PC) or telephone. This allows a user to access his files from anywhere he may be, via the Internet. Internet users increasingly use cloud services like Google Docs, Google Drive, Dropbox and iCloud. It is expected that cloud computing will become an important feature of the internet landscape (Koops et al., 2012). Cloud computing also offers interesting possibilities for businesses due to its flexible nature and the scalability[16] at relatively low cost.
In sum, the Internet is a digital environment where many people are active and which is used in many different ways. The boundary between the physical and digital is becoming increasingly vague as the Internet becomes more and more intertwined with 'normal' daily life. In the same vein, the distinc-

16 Scalability is a term that is used in the IT world, referring to the ease at which certain services or configurations can be scaled up.

tions between telephony and internet is getting smaller, and providers of these services increasingly focus on data streams. At the time of drafting the Data Retention Act, there still was a distinctly separate infrastructure for telephony and the Internet. At the level of detection and investigation there is also still a clear distinction in knowledge concerning the processing of telephone and internet traffic data. For the sake of clarity, we will maintain this distinction between the two throughout this report. In practice, however, this distinction is virtually non-existent.

2.3 Limitations of the retention directive

As mentioned earlier, technological advances and innovations have led to an increase in digital traces that are not covered by the Data Retention Act, as well as it being increasingly difficult for investigative services to retrieve data. This problem can best be explained using an everyday example.

> Mrs Smith gathers her belongings for the workday. After checking the latest news online on her laptop (via the Wi-Fi network at home), she closes her laptop and places it in her bag together with her mobile telephone. She then uses the home phone (landline) to make a quick phone call, after that she leaves the house. While waiting for the train at the station, she logs in to her network provider via a Wi-Fi hotspot with her smartphone, in order to check the news and her G-mail account. While doing so, she receives a 'WhatsApp' message to which she responds immediately. She then receives a call from her secretary and, whilst on the phone, proceeds to board the train. When Mrs Smith's conversation ends, she places the cell phone back in her bag and takes out her laptop to prepare for a meeting at work. For this she needs to access some information on the Internet and thus connects to the free Wi-Fi network on the train. Upon arrival, she closes her laptop and at work she logs in to the internal network of her employer and opens her work email.

This example illustrates a daily ritual for many people. It shows that people can communicate via many different channels and that this can proceed through many connection points. For example, connections to the network of telephone and internet service are not static and people can be connected with each other in many ways and from different locations. People also often use services supplied by different providers. Any given person may have a subscription with one particular provider for telephone and internet services delivered via the cable network, and at the same time have a subscription to another provider for the use of mobile telephone and internet. In addition, mobile calls can also be placed and internet can be accessed anonymously, through the prepaid services. For this, the user only needs to purchase a tele-

phone device with the required specifications and a SIM card with a credit for calling minutes or internet use. These technological innovations, and the resulting fragmentation of communication due to the use of various internet services, make it difficult to chart a person's distance communication. Moreover, not all traffic data generated by the varied use, is covered by the Dutch Data Retention Act, meaning that not all relevant data can be accessed.
At the request of the Ministry of Economic Affairs, Stratix Consulting examined how the interception of communications can be best guaranteed (Stratix Consulting, 2009; also see Koops et al., 2005). Interception here, refers to the securing of data for investigative purposes by claiming disclosure for user and traffic information, or by using a telephone or internet tap. One of the conclusions of this report was that the usefulness of the traffic data retrieved decreases with the technical developments and with developments in the phone and internet market. Many internet users have an email account with webmail services such as Hotmail, Gmail, or Yahoo, the provider of which is a foreign company. These providers or services offered do not fall under the Dutch Telecommunications Act and thus also are excluded from the Dutch Data Retention Act. As a result, relevant traffic data may not actually be stored for investigative purposes. In those cases, an international legal aid request has to be made in order to secure telecommunications traffic data, and even then, there is no guarantee of its availability. Communication with social media services such as Twitter, Windows Live Messenger[17], Hyves, Facebook, Ping, WhatsApp, and FaceTime often also proceeds through foreign servers. In all these cases, traffic data is (virtually) impossible to retrieve by Dutch investigative services. Moreover, it has become increasingly popular to use Voice over Internet Protocol (VoIP), calling via the Internet. Initially, internet calling was mainly used for international calls because it was a cheap way of calling. But as more and more phones have internet connections, it has become easier to make regular domestic calls over the Internet using a mobile phone. These services are not covered by the Data Retention Act and are often offered by providers falling outside the Dutch Retention Act (Preparatory Memorandum (House of representatives), 2007/ 08, 31 145, No. 9, p. 6). Telecommunications traffic data pertaining to cloud use can also only be accessed by Dutch authorities when the cloud service provider is a Dutch based public telecommunications service (for a detailed description see Koops et al., 2012). The Dutch authorities obviously cannot provide a basis upon which to retrieve data for investigative purposes when the network falls under the jurisdiction of another country.[18] In that case, the Dutch investigative services will have to request the legal assistance of their foreign colleagues and can only hope that the information is available.

17 Previously MSN.
18 *Explanatory Memorandum (House of representatives)*, 1989/90, 21 551, nr. 3, p. 12.

It also appears to be increasingly difficult to connect an identity to an internet user. An Internet Protocol (IP) address can change frequently, people can have access to services which allow them to surf the Internet anonymously and email addresses are easily changed, even by the user himself. Moreover, as mentioned earlier, many places offer free Wi-Fi network, for example at stations, on trains and in restaurants. Based on the traffic data stored within the Data Retention Act the user behind an IP address of a Wi-Fi network cannot be traced, because a user logged into a Wi-Fi network is assigned an IP address that is also used by all other persons that are logged on to the network at the same time.
The definitions of some telecommunications and network services in the Dutch Telecommunications Act, seem to leave some grey areas. The Data Retention Act covers 'public telecommunication services', in other words, 'services that are either wholly or primarily involved in the transfer of signals via electronic communication networks' and that are 'accessible to the general public' (Article 1.1 Telecommunications Act). The Data Retention Act contains no information about the volume or size of the network, which sometimes raises questions. The Independent Postal and Telecommunications Authority[19], supervisor of compliance with the laws and regulations in the field of postal and communication services and part of the Ministry of Economic Affairs, each year examines dozens of companies to determine whether or not they may be regarded as public electronic communications services and/or networks, and thus if they should be registered as such. The Telecom Agency uses the registry to enforce the 'tapping' obligation and Data Retention Directive. In 2003, the Independent Postal and Telecommunications Authority generally maintained that restaurant owners who wanted to offer their guests (only) internet access (for example via a Wi-Fi network) were not required to register with the Independent Postal and Telecommunications Authority (2010). In 2009, SURFnet, a provider of internet services for higher education and universities, filed a claim with the courts to determine whether or not it should be required to register with the Independent Postal and Telecommunications Authority, and as a result be obliged to store traffic data and facilitate network interception. At the time, the Independent Postal and Telecommunications Authority found SURFnet's target group to be so large that it deemed SURFnet as a public service, as defined by the Telecommunications Act. The Authority argued that the question was wether the users primarily wanted to communicate with each other, as in a closed company network, or were more interested in communicating outside of the network, through the Internet. The court did not agree and found that the criterion of 'communication with persons outside of the network' did not make the service or the network provided a public one. The court ruled that

19 Since April 1, 2013 the Independent Postal and Telecommunications Authority has pooled forces together with the Consumer Authority (*Consumentenautoriteit*) and the Dutch Competion Authority (*Nederlandse Mededingingsautoriteit, NMa*) to form the new supervisory organization Consumer and Market Authority (*Autoriteit Consument en Markt, ACM*).

the circle of people to whom SURFnet provided services can be regarded as a closed network as the people can be united in one target group, namely institutions of higher education and scientific research (ECLI: NL: RBROT: 2009: BH9324). Libraries, internet cafes, hotels, shopping centres, banks, restaurants and cafes are institutions that offer internet and/or telephone services to their customers. According to Independent Postal and Telecommunications Authority hotels that only offer Wi-Fi to private customers, cannot be considered as providers of public telecommunications services. For the other institutions it is not clear whether they are obliged to comply with the Data Retention Act. Neither the Telecom Agency, nor the Independent Postal and Telecommunications Authority have provided clarity in this matter.[20]
It is clear that due to the current phone usage, the rise of the smartphone, the general decrease in telephony, technological changes – such as the availability of Wi-Fi networks in public space – and the huge increase in internet services, only a small portion of the total historical telecommunications and internet traffic data is covered by the Dutch Telecommunications Act.

20 Also see www.ictrecht.nl/ictrecht/overzicht-bewaarplicht-wie-wel-en-wie-niet/ (consulted on April 4, 2013).

3 The legislative history and European regulation on the Data Retention Directive

On 3 May 2006 a European directive came into force, which aims to ensure that certain telecom and internet data is retained and thus are available for investigations into serious crimes. Accordingly, the member states of the European Union were required to convert this directive into legislation. The data retention period over which companies are required to store data can vary according to the directive, anywhere between a minimum of six months and two years. In the Netherlands, legislation on data retention has been ratified as per the terms of the directive. In Section 3.1 we discuss the background and purpose of the Data Retention Directive and the scope of member states who have also ratified legislation based on the Data Retention Directive. First we will discuss the implementation of the Directive and the development of legislation in the Netherlands. The main focus here lies on two central issues both in parliamentary debate and outside. On the one hand, there is the question concerning the purpose of the Data Retention Act in the context of strengthening criminal investigation. On the other hand, there is a concern for the protection of citizen privacy with regard to the (long term) storage of data.

3.1 The draft legislation

3.1.1 The nature of the data

The legislation covers so-called traffic and location data. Traffic data refers to data that is processed whilst transferring communications via an electronic communication network or the data generated in the billing thereof.[21] This may include the port number, the time, date, and duration of the communication, and the type of communication. Location data refers to information on the geographical position of the device used, e.g. the mobile phone. According to the Explanatory Memorandum (*Memorie van Toelichting, MvT*) of the draft legislation, witness statements about their whereabouts can, to a certain degree, be checked against the information provided by location data at the time of telephone communication.[22]
Pursuant to Article 13.4 Telecommunications Act, there already was a limited retention obligation to carry out a so-called file analysis. This served to enable a trace of user information if there was no registration data available with the provider, for example in the case of users of prepaid telephone service. To make this type of file analysis possible, providers were obliged to

21 The term 'traffic data' as used here, is defined in accordance with the Dutch Telecommunications Act; *Parliamentary Minutes II*, 2006/07, 31 145, nr. 3, p. 3.
22 *Parliamentary Minutes II*, 2006/07, 31 145, nr. 3, p. 9.

retain the necessary data for three months.[23] This is further detailed in the Data Telecommunications Recovery Decision (*Besluit vorderen gegevens telecommunicatie*).

The Data Telecommunications Recovery Decision[24] provides an exhaustive list of what is to be understood as traffic data. Specifically, the term data traffic covers user name and address, user numbers, name, address and number of the person with whom the user has connected or has attempted to connect with, the date and time at which the connection was established (or attempted) and terminated, and the duration of the connection, as well as the same information for connections attempted and/or made by others with the user of said provider. Furthermore, traffic data consists of information pertaining to the location of the network connection used or data on the geographical position of the user devices when a connection was made or attempted. This also includes the numbers of the device that the user uses or has used, the types of services that the user uses or has made use of as well as the associated data such as name and address of the person who pays the bill for public telecommunications services and telecommunications networks that the user has (had) available, if other than the user.

On the basis of a claim for traffic data disclosure, traffic data can only be obtained when indeed there is or was communication traffic. In other words, there must have been an actual or attempted connection made between an automated device (such as a telephone or computer) and another automated device. Mobile telephones on standby do generate location data, but cannot be exacted by investigative services under Article 126*n*/*u*/*zh* CCP because no communication traffic took place. Claims for the disclosure of location data for mobile telephones on standby can be made on the grounds of Article 126*ng*/*ug* jo. 126*ne*/*ue* CCP.

3.1.2 Retention periods

The Explanatory Memorandum following the Data Retention Directive distinguishes between telephony (fixed and mobile) and data relating to internet access, email and internet telephony. The Directive allows some leeway to make such a distinction and also allows for the possibility to maintain different mandatory retention periods for such categories. The draft legislation in the Netherlands initially made no distinctions in retention times for the two categories, but rather proposed a retention period of eighteen months for all retained data.

23 *Stb.* 2002, 31. In several European Union countries, prepaid telephone users are registered. The providers suggested file analysis as an alternative to registration of prepaid service purchase, which they deemed undesirable.

24 *Stb.* 2004, 394, last adapted *Stb.* 2006, 730.

This decision to initially propose an eighteen months retention period for all traffic data was based upon a number of considerations. Firstly, based on a study concerning the use of historical traffic data in criminal investigations, Mevis et al. (2005) found that a retention period of three months to generally be sufficient. However, for organized crime, fraud and serious, life threatening and violent crimes a longer retention period was considered desirable. This was also the case for cold cases and as well as for international requests for legal assistance. According to the Board of Police Commissioners (*Raad van Hoofdcommissarissen*) a longer retention period would be more effective, particularly in the case of organised crime and terrorism. In addition, for missing persons cases a shorter retention period could mean that traffic data are no longer available when the person is found at a later date. Furthermore, the explanatory memorandum points out that the use of traffic data in the investigation of crimes, both evidence of involvement can be found, as well as disconfirming evidence.
All in all it was thus decided to include a wider time margin in the retention period, to facilitate the use of that traffic data for more complex investigations, cold cases and international requests for legal assistance.
The reason why a retention period of twenty-four months was not elected had to do primarily with the fact that as long as there didn't appear to be a need for a longer retention period, the related financial costs and negative implications for the public's privacy outweighed the investigative interests. Because it was expected that the use of internet, and telephoning through internet, would increase in the future, it was decided that the same retention period should be maintained for both telephony as well as internet use.[25]
The Explanatory Memorandum suggested that the above considerations justified the importance of telecommunications data to be retained for eighteen months, a period considered proportionate to the privacy interests of the public as well as the costs of data retention.[26]

3.1.3 Protection of personal privacy

Traffic and location data can provide insights into individual behaviour. That is precisely the purpose of retrieving that data in the context of a criminal investigation. As a result, an infringement on the privacy of persons concerned takes place. The retention act means that, in compliance with the obligation to retain telecommunications data, this data has to remain stored even without the requirement for companies to retain the data as a business necessity for their own operations, such as billing.

25 This decision was also supported by a study carried out by Bureau Verdonk, Klooster & Associates BV concerning the costs of the implementation of the Data Retention Directive. For that study, various calculations were made using different retention periods (Boot, Van der Bosch, Vervaet & Varkevisser, 2006).
26 *Explanatory Memorandum (House of representatives)*, 2006/07, 31 145, nr. 3, p. 9.

There has been a legal basis on which law enforcement officials could request traffic information for some time now.[27] The Data Retention Act didn't change that. When there is a suspicion of a crime for which detention is authorized, when there is a reasonable suspicion that organized crimes are being planned or have been committed, or when there are indications of a terrorist offense, a claim for disclosure of traffic data can be made (Article 126*n*, 126*u* and 126*zh* CCP). What has changed as a result of the new legislation is the retention period for which data is to be kept. The Explanatory Memorandum of the draft legislation states that no further breach of citizens' privacy will thus result from the proposed legislation. There will merely be data available for more cases being investigated and/or prosecuted. This was also the main purpose of the bill.

The Explanatory Memorandum states that the data retention law meets the requirements in terms of privacy protection set by Article 8 of the European Convention of Human Rights and Article 10 of the Dutch Constitution. In a democratic society, any invasion of the privacy of citizens must be deemed a necessary one, meeting the requirements of proportionality and subsidiarity. In addition, the draft legislation sought to suggest measures against abuse or careless use of retained data. To this end, providers and telecommunications companies are required to take technical and organizational measures. Moreover, at the end of the data retention period, there is an obligation to destroy the retained data.

Customers of a provider or a communications service have the right to access the data pertaining to them that is being retained[28], although it is expected that the number of requests for access will be limited because the information about telephone calls is usually available on itemized bills. Monitoring compliance with the provisions of the bill is suggested to fall under the responsibility of the Minister of Economic Affairs and the Data Protection Agency.[29]

3.1.4 Notification

If a claim is made for disclosure of telecommunications traffic data[30] on the basis of the authority provided by the criminal code, e.g. the application of

27 The law distinguishes between information about a user of a communications service and communications traffic pertaining to that user (Article 126*n/u/zh* CCP), and the information pertaining to name, address, phone number and type of service of a user of communications services (Article.126*na/zi* CCP). These two categories are also referred to as traffic data and user data.

28 The exceptions to customer rights being when they conflict with the interest of the investigation or state security and when they conflict with the rights and protection of third parties (Article 43 Privacy Protection Act (*Wet Bescherming Persoongegevens, WBP*)). The provider is not permitted to provide information as to whether or not the General Intelligence and Security Service (*Algemene Inlichtingen- en Veiligheidsdienst, AIVD*) or the Military Intelligence and Security Service (*Militaire Inlichtingen- en Veiligheidsdienst, MIVD*) has made a claim for disclosure.

29 *Explanatory Memorandum (House of representatives)*, 2006/07, 31 145, nr. 3, p. 13 and 28.

30 On the basis of Articles 126*n*, 126*u*, and 126*z* CCP.

the prosecutor, Article 126*bb* CCP mandates a notification requirement.[31] This means that the public prosecutor is obliged to notify the person about whom information has been requested as soon as the interest of the investigation permits it, except in cases where such notice is not reasonably possible or in cases where persons – such as suspects – become aware of this through the court documents.
The Data Protection Agency initially noted that no attention was paid to a notification requirement in the draft legislation concerning the Data Retention Directive. However, it was noted in the Explanatory Memorandum that this was not necessary due to the fact that the Dutch Code of Criminal Procedure already included a notification requirement. In the meantime, new draft legislation proposes to abolish the notification requirement when it comes to claim traffic data.[32] The main reason for this draft legislation to propose an abolishment of the notification requirement concerns the inordinate amount of time that the execution thereof takes, burdening the Public Prosecutor Service with too much administrative work. To reduce this burden it is proposed to limit the notification obligation to the more substantial investigative powers: 'For the authority concerning serious infringements of the privacy of citizens, the notification will remain unchanged. For the authority posing a less serious infringement of privacy, the notification obligation will be abolished.'

Regarding the disclosure of traffic data, it is suggested that this authority poses a relatively light infringement of privacy and that using this authority is virtually standard investigative procedure in criminal investigations. 'Although the personal information can be obtained, the data itself is managed by third parties. Moreover, people know that others can access this information stored by the third parties.'

3.1.5 Consideration draft legislation by the Senate

The Senate paid extensive attention to the draft legislation for data retention. An expert-hearing was held on the subject and the exchange of views between members of the Senate and the Minister of Security and Justice led to a commitment to reduce the retention period for internet data (from one year to six months) on the part of the Minister. An amendment to this effect[33] became effective as of July 5, 2011. The written exchange of views with the Senate and the debate mainly focused on three subjects: the impact of the

31 The Special Investigative Powers Act (*Bijzondere Opsporingsbevoegdheden, BOB*) ensures that special investigative authority is legally regulated for a number of special circumstances. The use of this authority takes place without the person(s) under investigation being informed. There is a requirement (Article 126*bb* CCP.) that written notification be given after the fact.

32 Draft legislation proposing amendments to the Dutch Code of Criminal Procedure and the Dutch Code of Civil Procedure is designed to increase the effectivity of the police. See www.rijksoverheid.nl/documenten-en-publicaties/kamerstukken/2012/09/03/consulatieversie-conceptwetsvoorstel.html (consulted on May 1, 2013).

33 *Parliamentary Memorandum*, 2009/10, 32 185, nr. 2.

data retention proposal on the privacy of citizens, the effectiveness of data retention for investigative purposes, and the costs and burdens for private businesses, in the context of the assessment of utility, necessity and proportionality of the legislation and retention itself. Because the detailed consideration of the draft legislation in the senate provides a good insight into the pros and cons of the proposal (which were also discussed in parliament), we will discuss these more extensively in the next sections (3.1.6, 3.1.7, and 3.1.8).

3.1.6 Costs

In an explanatory response to the Senate, the Minister argued that the costs for the industry would be lower than suggested by deliberations in parliament. Where initially the additional costs of increasing retention periods from one to two years were calculated by Verdonck, Klooster & Associates BV[34] to be 14 million Euros, reducing the retention period by half a year would reduce the estimated costs to about 7 million Euros.[35]
As a result of the sharp decrease in costs of data storage (on the basis of the technology at the time) by shortening the retention period by six months, a reduction of – only – 4 million Euros could be achieved.[36] This is based on the Minister's assumption that a longer retention period would not involve proportionately more costs for, and burdens on, companies themselves.[37]
The effects of the retention directive on the costs for businesses did not play a (large) role in subsequent discussions, though this didn't hold true for the position of the smaller Internet Service Providers, whose costs for meeting the conditions concerning the protection and storage of data would be relatively higher.[38] As a result of the meeting of experts it was determined that smaller internet providers would face relatively higher costs, taking into consideration also the assumption that the number of claims for disclosure would be relatively low. In addition, these smaller providers differ greatly in the systems that they use which complicates storage standardization.[39] The Minister agreed to take the position of these smaller providers into consideration, and to maintain a dialogue with them. The Telecom Agency would be asked to take baseline measures to determine how the Data Retention Directive could be effectuated by the internet providers, and if it would be reasonable to expect them to take the necessary measure already in the start-up phase. This commitment was related to the fact that in September 2010 a European evaluation of the Data Retention Directive was provided. In the

34 Verdonck, Klooster & Associates BV, 'Research into the national implementation of the European Data Retention Directive'. October 9, 2006 (Boot, Van der Bosch, Vervaet & Varkevisser, 2006).
35 *Explanatory Memorandum (House of representatives)*, 2006/07, 31 145, nr. 3.
36 In earlier calculations it was assumed that each terabyte of storage would cost € 34,000, whereas later those costs fell to € 2,200 per terabyte. *Preparatory Memorandum (Senate)*, 2008/09, 31 145, C, p. 17.
37 *Minutes (Senate)*, 2008/09, 40, p. 1845.
38 Elaborated on in the decree Telecommunications Security (*Stb.* 2004, 394), last amended *Stb.* 2012, 615.
39 Verdonck, Klooster & Associates' research estimated there to be 255 small internet providers.

opinion of several members (parties) of parliament it was deemed preferable to evaluate before lumbering the internet providers with high costs.

3.1.7 *Effectivity of the Data Retention Directive*

Several groups in parliament questioned the effectiveness of the retention obligation, whereby a distinction can be made between telephony and other internet traffic data.
In general, it was deemed insufficiently substantiated that telephony data would contribute to criminal investigations. Specifically it was questioned whether a retention period of longer than six months was necessary, in light of the already existing six month minimum.
The Liberal Democratic Party in the Netherlands (*Volkspartij voor Vrijheid en Democratie, VVD*) inquired as to the number of criminal investigations in which traffic were used/required and to what extent they had proved to be decisive in the investigation. In support of the usefulness of traffic data, the Minister referred to the report by Mevis et al. (2005), as well as to the advice of the Board of Procurators General (*College van Procureurs-generaal, CvPG*), the police and some examples from case law showing the role traffic data had played. A longer retention period for traffic data would be especially useful for long term criminal investigations into organized crime, the international legal assistance requests and the investigations into missing person of whom later becomes apparent that foul play might be at hand. Moreover, for initially less complex cases, traffic data might prove useful to criminal investigations at a later time. In addition, investigations into cold cases has repeatedly been referred to in support of data retention. However, this argument in favour of data retention appears less valid because with cold cases, an existing file is re-opened. Generally re-opening a case would fall outside the retention period (of 1 or 2 years) anyway.
The data stored on the basis of the retention directive specifically provide insight into the relationship between perpetrator and victim, as well as into the nature and composition of criminal networks.[40] The Minister suggested that the sooner the traffic data is destroyed, the greater risk this would pose for not solving serious crimes.

The scepticism concerning the effectiveness of the use of internet data is greater among the members of the Senate. Various parties, including the experts, pointed out the enormity of the amount of data that would be stored under the Data Retention Directive, and that the greatest portion thereof would in fact be spam. Some estimates suggested that up to 95% of email traffic could be spam. Examples from abroad showed that, as the amount of data stored increased, so too would the risk of error in the data, the loss of

40 *Preparatory Memorandum (Senate)*, 2008/09, 31 145, C, p. 3.

large amounts of data.[41] In light of this, and the subsequent consequences, it was referred to a Newspaper article published in the NRC of May 21, 2008. This was a letter to the editor from of a number of professors who argued that innocent civilians would undoubtedly be the victims of inevitable errors. Such errors in data storage could lead to house searches and coercive measures on false grounds. It was noted by the Minister that as a possible measure against the significant proportion of spam in internet could be for the providers to filter spam out so that the transmission of the message to the recipient (end user) does not become effective. In those cases, internet providers are not obliged to store the data.[42]
Furthermore, it was pointed out that the Data Retention Directive could easily be bypassed through the use of telecommunications services not covered by the scope of the law[43], such as Hotmail, Gmail, Windows Live Messenger and Skype.[44] Additionally, it would be possible to conceal user identity, or even change user identity, through the use of certain software. The ease at which (stored) data can be manipulated could thus lead to a false sense of security. In response, the Minister replied that notwithstanding the use of such services, telecommunication via mobile phone and traditional email through Dutch providers would still be used.[45] The essence of the discussion was that only 'stupid crooks' would allow themselves to be caught as a result of the use of traffic data in criminal investigations, and that smart crooks would escape. This point was also brought up in the expert meeting. The Minister acknowledged that clever criminals could more easily bypass the workings of the Data Retention Directive and that the information which was brought up on this topic during the expert meeting, in fact was common knowledge amongst (smart) criminals. However, it was also suggested that clever crooks can be lazy and that wiretaps were thus still useful for the detection and investigation of crimes.[46] The objections to the Data Retention Directive pertained more to the relative utility of internet data than the use of traffic data relating to telephony.

3.1.8 *Privacy*

The Data Retention Act effects the privacy of citizens. This raises the question of how the Data Detention Act relates to Article 10 of the Dutch Constitution

41 *Preparatory Memorandum (Senate)*, 2007/08, 31 145, B, p. 7.
42 This solution was discussed by the experts during an informal meeting on January 22, 2009, concerning the implementation of the Data Retention Directive. *Preparatory Memorandum (Senate)*, 2008/09, 31 145, F, p. 3 (*NMvA*).
43 International email and internet services do not fall under the scope of the Data Retention Act, see chapter 2.
44 This point was also discussed in the House of Representatives.
45 *Preparatory Memorandum (Senate)*, 2008/09, 31 145, C, p. 30.
46 MP Franken (Christian Democratic Party, *CDA*) thought that the measure would only be useful in catching dumb criminals who don't use Skype or Hotmail. *Minutes (Senate)*, 2008/09, 39, p. 1808. In other words, as also follows from case law (*Bundesverfassungsgericht*) it was noted that the fact that criminals may find a way to bypass the law, no reason is to not pass the law.

and Article 8 of the European Convention on Human Rights[47] where civilian right to privacy is guaranteed. Storing traffic data could pose an infringement of privacy in two ways. Firstly, the mere storage of such data poses the risk that unauthorized persons – such as hackers – can obtain access to that data. The probability of such a violation of privacy would increase as the data retention period increases. A second, and other type of infringement occurs when law enforcement officials obtain access to retained data in the context of a criminal investigation. According to the European Convention on Human Rights, a limitation on the right to privacy is only allowed if it is provided for by law and is deemed necessary in a democratic society. Critical notes have been made by parliamentary parties, concerning the *necessity.* In the opinion of the Christian Democratic Party (*CDA*), given the doubts as to the actual utility of traffic data discussed above, the 'need' as expressed by law enforcement officials should better be referred to as a desire, as in 'nice to have', rather than considered as an actual necessity, as in 'a must'.[48] By extension it was questioned whether the law enforcement officials were not equally served by the traditional investigation methods. However, it was suggested by the Minister that traffic data were of such importance to the detection and investigation of serious crimes that the storage of that data thus met the necessity criterion.[49] The Minister argued that the infringement was not so much as a result of the storage of those data, as that it arose as a result of access to the stored traffic data.[50] Given the safeguards that surround data storage (as detailed in a draft security data), storing data for longer periods would pose no additional disadvantage to civilians.[51]
The Senate maintained a different opinion concerning the question whether the mere storage of data actually already forms a (severe) breach of privacy. In that regard, the Senate referred to case law (*Bundesverfassungsgericht*[52]), showing that the mere storage of data should be considered an infringement of personal privacy.

Access to and use of traffic is indisputably a curtailment of or – depending on the view – an infringement of the privacy of citizens. It is regulated in the Dutch Code of Criminal Procedure who has access to the stored data and under which conditions.[53] In case of suspicion of crimes for which pre-trial detention is permissible and/or when there is a reasonable suspicion that

47 Article 8, section 1: Everyone has the right to respect for his private and family life, his home and his correspondence. 2. There shall be no interference by a public authority with the exercise of this right except such as is in accordance with the law and is necessary in a democratic society in the interests of national security, public safety or the economic well-being of the country, for the prevention of disorder or crime, for the protection of health or morals, or for the protection of the rights and freedoms of others.
48 *Preparatory Memorandum (Senate)*, 2008/09, 31 145, C, p. 5.
49 *Preparatory Memorandum (Senate)*, 2008/09, 31 145, C, p. 11; *Preparatory Memorandum (Senate)*, 2008/09, 31 145, F, p. 12.
50 *Preparatory Memorandum (Senate)*, 2008/09, 31 145, C, p. 7.
51 *Preparatory Memorandum (Senate)*, 2008/09, 31 145, F, p. 6 (*NMvA*).
52 From March 11, 2008.
53 See the discussion in the Senate concerning the legal difference in definition between the infringement or the curtailing effect of a law.

crimes are being planned or committed, the public prosecutor can file a claim for disclosure of traffic data (Article 126*n*, 126*u* CCP).
An investigating officer may recover identifying data (Article 126*na*, 126*ua* CCP). In that case, suspicion of a crime or a reasonable suspicion of involvement in organized crime is sufficient. The data that can be retrieved are called user data. In case of indications of a terrorist offense, the public prosecutor is entitled to claim traffic data (Article 126*zh* CCP), and an investigating officer may recover user data (Article 126*zi* CCP). Furthermore, the public prosecutor can file a claim for disclosure from public and private organisations' databases, in an exploratory investigation concerning terrorist offenses (Article 126*hh* CCP). Although the legislation for retrieving the various data already existed independently before the Data Retention Act, the relation to the Data Retention Act is that the more data are stored, the more claims for disclosure can be filed. Therefore, a concern has been expressed that the storage of data would lead to a 'grab bag' in which law enforcement officials could freely operate. According to the Minister, there could be no question of this, now that access to and use of such information fall under the Dutch Code of Criminal Procedure.[54] Although data was stored for large groups of non-suspects, the Minister argued that a distinction should be made between the storage of and access to data.

The position that there was insufficient evidence supporting the need for data retention laws led a number of parliamentary parties to thus propose that retention periods be made as short as possible (i.e. the six month minimum mandated by the directive). The Minister advocated for a retention period of one year for both telephony and internet traffic data, in anticipation of future developments regarding the expected replacement of traditional telephony by internet telephony.[55]
Senate deliberations proved that a retention period of one year for internet data was untenable, given the opposition against it. As mentioned earlier, this resulted in the Minister agreeing to a legislative amendment of the Data Retentions Act after the ratification thereof by the Senate. The amendment would see the retention period for internet data reduced to six months and went into effect on July 5, 2011.[56] All in all, the retention period for telecommunications and internet data was reduced from eighteen months, in the initial to the bill, first to twelve months with the adoption of the amendment[57], and finally to six months as a result of the considerations in the Senate.

54 The data to be retrieved has been designated by a decree pertaining to the disclosure of telecommunications data (*Besluit vorderen gegevens telecommunicatie, Stb.* 2004, 394, last revised *Stb.* 2006, 730).
55 *Preparatory Memorandum (Senate)*, 2008/09, 31 145, C, p. 8.
56 *Preparatory Memorandum (House of representatives)*, 2009/10, 32 185, p. 5.
57 *Preparatory Memorandum (House of representatives)*, 2007/08, 31 145, p. 14.

As apparent from the above, the Dutch Senate was confronted with the obligation of meeting a European directive 'contre coeur'.[58] As a result, the Senate in particular has proactively and critically been following European developments pertaining data retention. For example, the Senate's objections to data retention have been voiced repeatedly in a critical response from the Justice and Home Affairs Council (*Raad Justitie en Binnenlandse Zaken, JBZ-raad*)[59] as well as in the permanent justice committee's report pertaining to the assessment of the European Commission's Data Retention Directive.[60]

These concerns include a brief yet convincing analysis of the absence of pressing social need, that not enough attention is paid to the proportionality of the measure, that the report insufficiently discusses the many possibilities to circumvent the data retention as well as the many remaining questions pertaining to the effectiveness of the directive.[61] Neither of the committees were convinced that the used case positions support the usefulness of data retention for criminal investigations cited in the report. In particular, case positions put forward by the Netherlands, where the evaluation unfairly spoke of providing evidence. This, as was the response of the committee, despite the fact that retained data[62] can only be used as an investigative means that may give rise to evidence. However, this appears to be incorrect. In Chapter 6 of this report examples are given of court rulings in which traffic and location data are used as direct evidence (as opposed to giving rise to evidence).

3.2 The European guidelines

European background for the retention of data

According to the Data Retention Directive, member states are required to compel providers of publicly available electronic communications services or of public communications networks to store traffic and location data between six months and two years, for the purpose of the investigation and prosecution of serious crime.[63]

As it is, providers of (tele)communications services process e.g. personal data relating to the communication for invoicing purposes. The source, date and time, duration, location and nature of the communication can be derived

58 *Minutes (Senate)* July 7, 2009, 40-1839 e.v. It was noted by MP Franken of the Christian Democratic Party, that certain arguments that had been exchanged were in fact not in line with the European directive nor did they fall under the scope of the law. For this reason it was suggested that one had to go to Brussels again to have the directive amended.

59 The council formation for Home Affairs and Justice.

60 Evaluation of data retention directive (*Evaluatie van de richtlijn gegevensbewaring*), a report of the Committee to the Council and the European Parliament, (Richtlijn 2006/24/EG). *Preparatory Memorandum (Senate)*, 2010/11, 32 797, A.

61 *Preparatory Memorandum (Senate)*, 2010/11, 32 797, A, p. 2.

62 This probably refers to retained traffic data.

63 The directive applies to telephony via a fixed network, a mobile network, internet access, email over the Internet and internet telephony (Article 1 section 2; Article 3 section 2; and Article 5).

from these data. Under Directive 2002/58/EC concerning privacy and electronic communications, such traffic data should in principle be erased or made anonymous when it is no longer necessary for the service and billing.[64] Before the directive came into effect, under certain conditions authorities could request providers access to such data, in the interest of law enforcement. For example, information about which subscribers used a particular IP address, or where a mobile phone was at any given time.
The use of data was first regulated at EU level by Directive 97/66/EC. This directive gave the opportunity (but not the obligation) to take legal measures in order to – among other things – secure public safety, protect national security and to facilitate the enforcement of criminal law. The 'e-Privacy Directive' allows member states to derogate from the principle of confidentiality of communications and indicates under what conditions the storage of, access to and use of data for law enforcement purposes is permitted. Under Article 15, paragraph 1 of Directive 97/66/EC, member states may restrict privacy rights and duties, for example, by storing data for a certain period, in the case of a democratic society deeming it necessary, appropriate and proportionate and in the interest of national security, i.e. state security, national defence, and public security or for the prevention, investigation, detection and prosecution of criminal offenses or unauthorized use of electronic communication.

As a result of the legislation adopted by different member states under Directive 97/66/EC and e-Privacy directive, the need arose for providers in those countries to acquire equipment and employ staff in order to be able to comply with law enforcement authorities' requests for information. This, despite there being no such requirement in other countries, which in turn led to a distortion of the national market. In addition, certain developments resulted in Internet Service Providers storing less traffic and location data. As a result, less data was available for use by law enforcement authorities. In the context of the terrorist attacks in Madrid in 2004 and in London in 2005, the EU directive on the retention of telecommunications data was adopted, which obliged all member states to retain communications data, thus making them available for the investigation, detection and prosecution of serious crimes. The Directive amended Article 15, section 1, of the e-Privacy Directive with a new provision to the effect that Article 15, section 1, would not apply to data retained under the Data Retention Directive.[65] Neither the Directive nor in the e-Privacy Directive includes a definition of 'serious crime'.

64 Unless the user gives permission to the provider to retain these data.
65 Article 11 of the directive states that in Article 1*bis*. Section 1 is not applicable to the data specifically required by Directive 2006/24/EC of the European Parliament and of the Council of 15 March 2006 on the retention of data generated or processed in connection with the provision of publicly available electronic communications services or of public communications networks to be retained for the purposes referred to in Article 1(1) of that Directive.

3.2.1 Retained data

Article 5 of the Directive lists the categories of data to store:
- the source of a communication;
- the destination of a communication;
- the date, time and duration of a communication;
- the type of communication;
- the communication equipment or the presumed communication of users;
- the location of mobile communication equipment.

The Directive also covers unsuccessful call attempts.[66] No data may be kept from which the content of the communication can be revealed.[67] With the exception of Belgium, the implementing legislation in twenty-one countries serves each of these categories of data.[68]

3.2.2 Ratification of the Data Retention Directive in the European Union

European Union member states were obliged to ratify the directive before September 15, 2007, though an extension was given until March 15, 2009 for the internet data retention. In amongst others Austria and Sweden, the Directive has not yet been converted into legislation. The constitutional courts of the Czech Republic, Germany and Romania have annulled legislation derived from the data retention directive.[69] The legal basis of the Directive was challenged in vain by Ireland before the European Court of Justice. The reasoning was that the main objective of the Directive would be the investigation, detection and prosecution of serious crime, and thus, fell under the third pillar of the European Union, namely police and judicial cooperation in criminal matters. However, the Court ruled that what the directive regulates is independent of the performance of any form of police and judicial cooperation in criminal matters and that the directive in its essence deals with the activities of service providers in the relevant sector of the internal market.[70] In the countries where the Data Retention Directive has been converted into legislation, the duration of the retention period chosen differs greatly. The storage bandwidth provided by the directive, namely retention between 6 and 24 months, has been fully utilized. Some countries, including the Netherlands, have opted for different retention periods, depending on the type of

66 Article 3, section 2.
67 This also applies to searches (server logs) because they shouldn't be regarded as traffic data but as holding content.
68 Report of the commission to the Council and the European Parliament, Evaluation of the Data Retention Directive (Directive 2006/24/EG, Publication of the European Union, April 13, 2006).
69 Ruling nr. 1258 of October 8, 2009 of the Romanian Constitutional Court, Romanian Official Gazette, nr. 789 of November 23, 2009; case of the *Bundesverfassungsgericht* 1 BvR 256-08 of March 2010 concerning the provisions of chapter 97, point 3 and 4 of law nr. 127-2005 concerning electronic communication and to amending related decrees, and decree nr 485-2005 concerning data retention and transmission to authorized authorities.
70 Court of Justice of the European Union, February 10, 2009, nr C-301/06 (Irish/European Parliament and the Council of the European Union).

data being stored. Roughly speaking, a distinction can be made between data relating to traffic via internet and telephone traffic data. In all these cases, the retention period for internet is shorter than that of traffic data relating to fixed and mobile telephony.

To give an idea of the retention periods for the different countries they have been summed here from shortest to longest retention periods.[71] The two countries with the shortest retention periods are Cyprus and Luxembourg, holding the minimum retention period of six months. Three countries, namely Malta, the Netherlands and Slovakia, apply a retention period of one year for fixed and mobile telephony and six months for internet data. Eleven countries have opted for a retention period of one year: Belgium, Bulgaria, Denmark, Estonia, Finland, France, Greece, Hungary,[72] Spain, Portugal and the United Kingdom.[73] Slovenia and Latvia have chosen respectively fourteen (eight months for internet data) and eighteen months. Italy, Ireland and Poland use the maximum storage period of two years.
This overview shows that more than half of the countries (14 out of 23 countries who have implemented the directive) maintain a retention period of one year for fixed and mobile telephony data. With regard to the storage of internet data, the ratio is slightly different: six countries opt for a retention period of six months and eleven countries for a period of one year.[74]

It has already been noted that the term 'serious crime' is not defined in the guidelines. This can be seen in the various grounds for legislation in the various member states pertaining to the retention and access to the retained data for criminal investigation purposes. In ten countries serious crimes are defined as crimes for which a minimum prison sentence holds – or merely a prison sentence – and where is referred to a list of serious crimes elsewhere in national law.[75]
In eight countries[76] data can be stored for purposes of investigation of all offenses as well as for crime prevention in general or on general grounds of national security.[77] In four countries, the term 'serious crime' or 'a serious offense' is included in the legislation without further definition. Some countries have gone further than required by the directive, in granting access to

71 Report from the Commission to the Council and the European Parliament, Evaluation of the Data Retention Directive (Directive 2006/24/EC, Official Journal of the European Union, April 13, 2006).
72 A retention period of six months is maintained for connection attempts.
73 In Finland and the United Kingdom smaller providers are not obliged to retain data because the costs for them are disproportional to the gains in terms of criminal procedure.
74 Member states 'can under specific circumstances justify a limited extension of their maximum retention periods'. A submission for extension must be submitted to the Commission, who in turn can grant or deny an extension. Thus, the maximum retention period can be extended, but not shortened.
75 Bulgaria, Estonia, Ireland, Greece, Spain, Lithuania, Luxembourg, Hungaria, the Netherlands and Finland (see: Evaluation of the European Commission on the European Data Retention Directive (COM (2011) 225 final), Table 1, pp. 6-8).
76 The report by the European Commission is not precise here, because there is no clear distinction made between purpose limitation regarding the storage of data and the access to data for criminal procedural purposes.
77 Belgium, Bulgaria, Denmark, Greece, Estonia, Ireland, Spain, France.

data. For example these countries allow access to data for the prevention and combating of crime in general, whereas the directive specifies 'serious crime'. The wording of the e-Privacy Directive permits this.
As for the duration of the retention period, it holds that the harmonization with the EU legislation has been sought, though has only been achieved to a limited degree. In the opinion of the Commission the lack of harmonization is the result of differing costs for providers in different countries, because differences in the purpose of data retention yield differences between the member states, in the number of requests for data. In addition, this could mean that there is an insufficient degree of predictability that must be required of a legal measure which limits privacy.
The access to the data by others than the providers themselves, has also been organized differently in member states. In all member states, both the police and the Public Prosecutor are considered legitimate authority to access the data. In eleven countries, however, permission from a judge is also required. In fourteen member states national security agencies may also request data. There are also countries where the same applies to customs authorities or tax authorities.

3.2.3 *Evaluation of the directive*

In accordance with Article 4, the European Commission has evaluated the application of the Data Retention Directive.[78] The purpose thereof was to determine whether or not the directive needed to be adapted. In addition, the report addresses the impact of the Data Retention Directive on fundamental rights within the context of the general criticism concerning data retention. Furthermore, the evaluation focuses on concerns about the use of anonymous SIM cards in criminal practices.
In order to monitor the implementation of the directive, member states are required to annually provide the Commission with statistical information relating to the cases where a claim for data disclosure has been made, the age of the data[79], and when disclosure requests have been granted. The statistics provided to the Commission differed in scope and detail of information, ranging from statistics with information concerning communication type and age of the data, to statistics without further classification. A total of nineteen EU member states provided statistics on the number of claims for disclosure in 2009 and/or 2008. Of those nineteen states, several had either not yet ratified the Data Retention Directive or had annulled legislation pertaining to it. Seven member states who had ratified the directive did not provide the required statistics.

78 Report of the commission to the Council and the European Parliament, Evaluation of the Data Retention Directive (Directive 2006/24/EG, Publication of the European Union, April 13, 2006).
79 That is to say, the amount of time elapsed between the initial date of retention and the date when a claim for disclosure is made.

A comparison of the statistics between member states had proved extremely difficult, even for the bare numerical counts of the number of requests for data disclosure. This is due to the difference in the ways in which member states register these data (see also Odinot et al., 2012, p. 295). Some countries record the number of requests per person, whereas other countries record the number of claims for disclosure made per telephone number or IP address. In countries such as the Czech Republic, Latvia and Poland, any request for information is submitted to different providers so that each request can end up being counted several times. It remains unclear from the evaluation report whether all countries have provided information as to the number of claims for disclosure, or if they have only provided information for *granted* claims for disclosure. In theory it is possible that unanswered requests in one country may be included in the statistics (as is the case in the Netherlands, The Czech Republic, Latvia and Poland for example) but not in other countries. The difference in number of claims for disclosure varies greatly between countries, ranging from less than 100 claims in Cyprus to more than one million in Poland, likely owing to the differences in statistical registration method.

In the evaluation, the Commission notes that the member states started implementing the Data Retention Directive later than expected, particularly with regard to internet data.[80] As a result, several countries have failed to comply with the obligation to provide the indicated statistical information to the Commission.[81] In 2010, the Commission requested member states to provide information as to the extent to which retained traffic data had been used for law enforcement. Ten countries have complied with this request.

The member states that actually completed a questionnaire concerning the usefulness of the data retention, expressed that the data retention was at least valuable and sometimes indispensable for preventing and combating crime, the protection of victims and the acquittal of innocent persons in criminal proceedings. Czech Republic found the retention of data to be ‘absolutely indispensable in many cases’. Hungary described it as ‘indispensable for the activities of law enforcement’, Slovenia argued that not retaining data would ‘paralyse the operation of law enforcement agencies’ and police from the United Kingdom indicated that the availability of traffic data was ‘absolutely crucial for the investigation into the threat of terrorism and serious crime’. According to the member states, the use of retained data enabled them to trace witnesses who would otherwise have remained unknown, and provided evidence or indications of complicity in criminal offences. Some member states also claimed that thanks to the use of retained data, the innocence of suspects of crimes could also be proved, without any need to use other, more invasive methods such as interception and searches. Based on these findings,

80 Member states were obliged to have ratified the directive before September 15, 2007. An extension until March 15, 2009 was given for the ratification of the Data Retention Directive pertaining to internet traffic data.

81 Mentioned in Article 10 of the directive. Nine member states were able to provide all the required information for the years 2008 and 2009.

the evaluation of the directive concluded that on the whole, data retention has been shown to be a valuable tool for law enforcement and criminal justice systems. The evaluation further states that the directive has not resulted in the desired harmonization when it comes to purpose, limitation, or retention periods of data. In order to ensure the privacy and protection of personal data, the report concludes that a high degree of security in the storage and use of traffic and location data should be strived for.

It has already been noted that the term 'serious crime' is not defined in the guidelines. This can be seen in ways that the legislation of the various member states have embedded access to retained data for criminal investigation in the law.

In addition to the Data Retention Directive and the obligations arising from that, we point out here yet another EU regulation relating to the freezing of data (also called 'quick freeze').[82] This enables authorities, upon the detection of a serious crime (attempt), to make an immediate claim to providers to retain any traffic data hence forth.

In the discussion about the pros and cons of data retention, it has been argued that the 'freezing' of data, when there is a specific reason (e.g. a committed serious crime), can have a much greater effect than overall retention of non-suspect citizens. The evaluation has not been able to establish to what extent the method of freezing data would bear fruit.

With respect to data retention's value for criminal investigation, incidental examples of concrete cases have been provided by several countries, showing how telephone or internet data have contributed to the solution of crime. For example, using IP addresses has been a means for investigating an international paedophile network. Statistical information for 2008 has revealed that in about 90% of cases data has been requested over a period of less than six months. Nevertheless, according to the member states, data from a date earlier than six months ago may be crucial for the investigation of crimes in certain circumstances. Although no statistical information is available about the value of data requested for evidence, the use thereof is asserted to constitute an integral part of the investigation.

3.3 Conclusion

Opinions on the need and the desirability of data retention have varied both between the European member states as well as within the Dutch Parliament. The main points of concern in the discussion pertained to the cost of implementation, the invasion of privacy of citizens and the extent to which the use of traffic data was useful and effective in the investigation and prosecution of crimes. This has led to differences between the member states, regarding

82 Article 16 of the Cybercrime Treaty.

retention periods and use of data, in the implementation of the Data Retention Directive. In this sense, the intended harmonization has not been achieved. The Netherlands ultimately opted for a retention period of one year for telephony data and six months for internet data. Although some empirical studies formed a base for the discussions on the effectiveness of traffic data for the investigation and prosecution of crimes (see eg. Mevis et al., 2005), ultimately this only provided a limited basis for the final decisions concerning the length of the retention periods. There also was Dutch criticism concerning the limitations of the cases presented in the European Commission's evaluation. The lack of empirical support has influenced the arguments of both proponents and opponents of (an extension of) the Data Retention Directive.

4 The retention and securing of data in practice

This chapter discusses the role of providers and how they deal with the Telecom Data Retention Act, and the role of regulators: the Telecom Agency and Data Protection Agency. It describes how the monitoring of data retention is regulated in practice. However, research on how the regulators carry out their monitoring duties, was not part of this study.
Employees in key positions of two major providers, one small provider and an internet hosting company were interviewed. Interviews were also conducted at the Telecom Agency and Data Protection Agency. First, we will describe the role of supervisors and subsequently we will address the role of telecommunications service providers.

4.1 The regulatory authorities

The Data Retention Act not only contains rules about the type of data to be stored, the retention periods of these data and the circumstances under which these data may be used, it also describes how to monitor compliance. It involves the preservation of sensitive data and it is thus important that the retained data are not used by other agencies or for purposes other than those for which they are stored. The monitoring of compliance with the rules is in the hands of the Telecom Agency, which operates as an independent regulator and enforces the Act. The Telecom Agency is part of the Ministry of Economic Affairs and is directly accountable to the Minister of Economic Affairs.[83] This chapter provides a picture of how the supervision is organized. The Data Retention Act contains safeguards to protect the privacy of telecommunication service users. In addition, the Data Protection Agency has a supervisory role. The Data Protection Agency is an independent administrative body, set up by law, to ensure that individuals and organizations (amongst others) comply with the Privacy Protection Act (*Wet Bescherming Persoongegevens, WBP*). The Data Protection Agency oversees all legislation involving the storage, processing or use of personal data. The Telecom Agency collaborates with the Data Protection Agency to ensure that no improper use is made of retained data, regarding compliance with the Data Retention Act. The Data Protection Agency does not actually conduct measurements, nor does it independently investigate compliance with the Data Retention Act. When an incident or a violation occurs, the Data Protection Agency reports this to the Telecom Agency, who in turn takes action. This is stipulated in an agreement between the Telecom Agency and Data Protection Agency.

83 See www.agentschaptelecom.nl/sites/default/files/toezichtsarrangenment-dataretentie.pdf (consulted on July 1, 2013). Contrary to what is stated in the supervision arrangement, in the mean time accountability is directly to the Minister as opposed to the under secretary.

The Telecom Agency ensures that only location and traffic data are stored, as opposed to e.g. the content of contacts or conversations. Further the telecom Agency ensures that the stored data does not fall into the wrong hands and that the stored data is adequately destroyed after the expiry of the retention period.
The Telecom Agency checks whether providers have taken sufficient measures to ensure protection of the retained data. Providers are obliged to describe their security measures regarding the retained data in a security protocol. This security protocol must also meet legal requirements. The Telecom Agency oversees the security and control processes. If the Agency suspects that the rules are insufficiently obliged, it proceeds with further inspection of the provider. In the case of infringement, action is undertaken and penalties can be imposed.
Every three years, the Telecom Agency writes a supervision plan that has a validity of three years. The plan is designed to inform the relevant parties how the monitoring process of the Telecom Agency takes place and it addresses specific priorities requiring attention in that particular period. The priorities are determined by (recent) legislation, allowing the various parties to see how the regulators will check and control compliance. The supervision plan also addresses maximum possible fines for offenses.
In practice, supervision is organized in an annual 'supervision cycle' during which all providers known to the Telecom Agency are approached by mail. The providers receive a survey form that addresses the focal points of the supervisory plan. In the questionnaire, providers are asked to indicate the current status pertaining to, amongst others, data retention legislation.[84]
The Research and Documentation Center (*Wetenschappelijk Onderzoek- en Documentatiecentrum, WODC*) researchers asked the Telecom Agency employees if they felt that, as supervisors, they were getting the right information from the telecommunications service providers' self-reports, concerning the in-company status and regulation. The interviewees suggested that they had the impression that they were being properly informed by the companies, and indicated that they took into account the possibility of socially desirable answers (provided by interviewees) in assessing the questionnaires. Based on the responses and data from the questionnaire completed by the providers, the Telecom Agency reviews the biggest potential risks. These are closely supervised. Further, the largest relevant providers are selected. Next, the Telecom Agency plans inspection visits to assess the actual situation, giving highest priority to those companies posing the highest risk. An important factor in determining and assessing the potential risks of the different providers is the size of the company. The interviewees at the Telecom Agency explained that company size provides an indication of the number of customers and therefore the relevance of the data stored for the

84 Also see www.agentschaptelecom.nl/sites/default/files/eindrapport-nulmeting-wetbewaarplicht-telecommunicationsgegevens.pdf (consulted on July 26, 2013).

criminal investigation purposes. The major telecom providers generally also take responsibility for the storage of traffic and location data of smaller providers who rent these parts of the network. According to the Telecom Agency interviewees, and according to the major providers, the monitoring and supervision of the smaller telecom providers is thus indirectly guaranteed. Smaller providers then are merely responsible for the execution of the Data Retention Act pertaining to customer data. Thus the larger providers are inspected sooner than smaller providers.

> *'[...]. If you want to achieve results, it is important that it is well organized with the major parties, because it is there that you have the greatest impact.'* – Telecom Agency

The Independent Postal and Telecommunications Authority[85] uses a class system that classifies providers by size. This is measured on the basis of provider turnover. The classifications are: small, 0-2 million Euros; medium, 2-20 million Euros, and large, 20 million or more Euros. A small provider suggested, however, that this was not an airtight approach. This particular provider offers telephony services over the internet at very low cost, and its turnover can be considered small compared to the major providers, but the provider's number of customers has been increasing for years, while sales only show modest increases due to the relatively low revenue per customer. In contrast, the larger providers offer more complete services at a higher cost, yielding a relatively higher turnover.

In addition to the annual survey addressing focal points and priorities for supervision plans, the Telecom Agency aims to visit all approximately 600 registered telecom service providers once every four years. Focal points of the current supervision plan are extra emphasis on security, storage and destruction of data. Regarding security, this refers to both the physical access to the building as well as to the security of the ICT environment. Regarding storage, verification that the correct data set is saved is a priority. As to the destruction of data, focal point is whether or not this is properly regulated after the expiration of the retention period.[86]

On the basis of the baseline measurement of the Telecommunications Data Retention Act (2010), or the 'State of the Ether' annual report (2012), it is impossible to determine whether or not sanctions have been imposed on parties who didn't meet the various requirements of the Data Retention Act. The baseline of the Act retention Telecommunications Data (2010) and of the annual report 'The State of the Ether' (2012) cannot be inferred if sanctions

85 Since April 1, 2013 the Independent Postal and Telecommunications Authority has pooled forces together with the Consumer Authority (*Consumentenautoriteit*) and the Dutch Competion Authority (*Nederlandse Mededingingsautoriteit, NMa*) to form the new supervisory organization Consumer and Market Authority (*Autoriteit Consument en Markt, ACM*).

86 Also see www.agentschaptelecom.nl/sites/default/files/toezichtsarrangenment-dataretentie.pdf (consulted on July 1, 2013).

are imposed on parties that do not meet the different requirements of the data retention law.[87]

4.2 The providers

To understand how providers deal with the obligations stipulated by the Data Retention Act, four providers were addressed: two small and two large. The major suppliers approached were large companies that provide various products and services, including internet and telephone services, for both the business and consumer markets. In the Netherlands, there are six major providers who together hold the telecommunications network. These companies rent out their networks to smaller telecom providers who in turn sell services to consumers. When a small provider rents a part of a telephony network from a large company, the large company takes responsibility for the storage of traffic and location data. The use of the storage capacity and the systems required to retain data are furnished by the major supplier, who includes it in the rental costs. The small provider is still responsible for the proper and safe storage of the names and addresses of its customers, as per requirement of the Data Retention Act.
Before the ratification of the Data Retention Act, retention periods varied between providers. One interviewee indicated that, before the law came into force in 2009, his company maintained a three month retention period. Another provider interviewed for this study indicated that the company stored data longer in 2009. It reduced its retention period from two years to one year. Having said that, the implementation of the Data Retention Act in 2009 didn't come as a surprise, and was anticipated by the major providers who were interviewed by us.

> *'Though we had quite some work as a result, the implementation time was long. During the amendment of the legislation, there was already consultation with the parties before it was introduced. In 2006 we already started the organization of the database. [...]. You do not want to be surprised by a new law.'* – Provider

After the implementation of the Data Retention Act, both providers interviewed had to adapt their retention periods as well as the content of what was being stored. Although adapting the content or lengthening or shortening retention periods may sound like simple procedures, in practice they pose a complex technical problem.

87 www.agentschaptelecom.nl/sites/default/files/eindrapport-nulmeting-wetbewaarplicht-telecommunications gegevens.pdf; www.agentschaptelecom.nl/sites/default/files/staatvdether_2012_digitaal.pdf.

'Data must be stored separately in a certain way. [...]. You must have the ability to perform searches in the retention environment. You have to cluster information in a different way than you need for your own billing process. [...]. Apart from just having the data, you have to group them, bringing data from different systems together in order to meet requests for data disclosure.' – Provider

Despite the long lead time, the implementation of the Data Retention Act was an extensive project for both providers interviewed. The two providers had different experiences concerning the costs incurred for the implementation of the retention act. Where one provider said to have invested millions, the other provider said to have built the necessary infrastructure itself, thereby keeping costs low. Both providers interviewed indicated not to have been compensated in any way by the government for the costs incurred.

Indeed, the government provided no compensation for the investment needed to set up for the retention of the databases. Personnel required to provide requested data is reimbursed by the government.
An agreement was signed with the major providers for the overall service of the providers to the government on the basis of the Criminal Code, the Telecommunications Act, the Law on the Intelligence and Security Services Act 2002, the Police Act 2012, the General Administrative Law Act, the general law on state taxes and the underlying regulations. This agreement also covers the provision of information under the Data Retention Act, but because the content of the agreement is not public, it is not known whether this is sufficient.

The data, which providers need to retain since 2009, are stored in a separate database that is located in a secure environment and is accessible only to a limited number of employees. This is in contrast to the normal database that each provider uses for its administration and that is also used by call centres. Each provider manages its own database containing the data as prescribed by the Telecommunications Act. Because each provider fills, manages and secures its own database, the storage of traffic and location data from the phone and IP traffic is decentralized in the Netherlands. The Data Protection Agency is pleased with this.

'We have always opted for decentralized storage and even encouraged it. In our experience, especially the large telecommunications companies and Internet Servers Providers are quite skilled in maintaining adequate security.' – Data Protection Agency

The law states that after the indicated duration of processing the data must be deleted or anonymised. Anonymising data means that the data is completely and irreversibly stripped of any identifying characteristics. For the two major providers who were interviewed for this study, data stored under the Data Retention Act is automatically destroyed after the expiration of the retention period. Other data in the records are anonymised and used for business purposes, unless the customer has indicated otherwise.

> *'The directive suggest that it is preferable not to have additional databases on top of the normal database providers use for billing and the like. For privacy reasons. [...]. But this is what major providers do and therefore the destruction of data is done really well. A script is placed over it, and an additional script over that, so that the data is irreversibly destroyed. This is in accordance with the law.'* – Telecom Agency

The researchers wondered whether the data stored for retention and kept in the specially built databases were perhaps also stored in a different way in another part of the provider's systems. In that case, the data would appear to be destroyed in accordance with the law, but in fact still exist elsewhere in the system. Further inquiry at the Telecom Agency and the providers showed that this is only partially true. The provider inserts the customer's name and address and the traffic data into a special database for traffic data retention. If a customer switches to another provider, the customer data in this database will be destroyed by the provider when the retention period has expired. However, data pertaining to a customer's name and address are also stored together with the data required for billing purposes in another database. According to the Telecommunications Act, the relationship with these data ends one year after the customer has ended its use of a particular provider. If the company wants to keep the data longer, these data fall under the Data Protection Act, which states that if a provider wants to keep certain data beyond the expiration of the retention period, for example for business purposes, the provider is required to anonymize it. In addition, providers are obligated to inform their customers of the nature of the data, and the retention period thereof, stored for the purpose of billing. This also applies to data stored for other purposes, only in that case, providers must obtain explicit prior subscriber authorization, which incidentally is often only done passively by providers. The customer must actively indicate that he does not want his data to be used for (other) business purposes.

Location details and received phone calls are, on the other hand, not necessary for billing and are thus not stored in the normal provider database. This also holds true for IP data which often have little or nothing to do with the financial settlement with customers. This data is only stored by mandate of the Data Retention Act and are automatically destroyed by the companies interviewed after the expiration of the retention period.

The small provider interviewed for this study, revealed that the method and the manner in which is dealt with the obligations arising from the Data Retention Act, is different to that of the large providers. This provider only recently became active in dealing with the retention periods because the amount of data managed became too large. For this provider it was primarily a practical consideration to reduce the retention period to a year. However, there is no separate database for managing traffic and location data. When a claim for disclosure is received, an employee manually extracts the requested data from the system. This takes about half an hour per application to complete.

> *'We've kept growing in recent years and we still are, so there is more and more traffic. It is because of this growth that we only now need to limit the telecom data. We are now actively bringing it back to a year. We've been allowed several years to do this. At times it was useful for the statistics. Or when a new customer registered with us we thought, "Hey, perhaps this is a returning customer who left us earlier with payment defaults". But now the mountain of data has gotten so big that we actively take steps to reduce it to 1 year.'* – Provider

Costs are associated with the compliance to the legal obligation to retain data for a year and subsequently destroy it, for which the provider is not reimbursed by the government. The implementation of this obligation is an even bigger problem for the smaller providers. As mentioned earlier, the Dutch government has entered into an agreement with the major Dutch providers. This agreement regulates compensation for the staffing required in order to disclose data stored under the law. This particular small provider does not belong to the larger providers and thus receives no compensation. According to the interviewee, the activities associated with complying with the legal obligations place too much pressure on the company, which has only a handful of employees.

> *'The costs [of storage] are limited, under € 5,000 per year. You have a server, rental of additional server space, and extra man hours to keep that thing going. We do not want pity. But we have X number of employees, which is a good amount to make it profitable, but technical developments go very fast so you have to invest a lot of energy to keep up. That [data retention] is something we can do without, because we already have a full technical agenda and there are a litany of things we all would like to do and then this [data retention] must be fit in between.'* – Provider

The situation sketched by both this small provider as well as the major providers corresponds with the image that the Telecom Agency has of the implementation of data retention. In 2010, the Telecom Agency conducted a baseline measure for the implementation of the Retention Act by Internet Service

Providers.[88] This study showed that the implementation of the data retention and compliance with retention laws, was in good order with the major providers. Moreover, this study showed that smaller providers are less strict to follow the letter of the law in their regulation of the obligation to retain data, and that the destruction but also the retention of data is not always done correctly. Contact between providers and the Telecom Agency shows that especially the interplay between data retention and use of traffic and location data for business purposes is complex. Stored traffic and location data are often not destroyed in time, because they are used for business purposes. The question is whether these providers are adequately informed of the obligations associated with the processing of traffic and location data for business purposes. This is being investigated further by the Telecom Agency.
In addition, the smaller providers are often unfamiliar with security plans or standards. It must be noted that, according to the experts interviewed in the Telecom Agency, it is possible to meet the security requirements without the use of a standardised security method, and thus these findings must be seen in perspective: the number of times small providers receive a claim for data disclosure is relatively small. The small provider interviewed for the current study received approximately ten data requests per year.

A fourth provider interviewed for this study revealed another situation. This is a small provider of web hosting services, where customers run their websites on the company servers. It appeared that the systems were designed and utilized in such a way that no data was stored. The company is run by persons holding privacy in high regard, and as such advertise that their company adheres to a so-called 'privacy-by-design' principle. The owners of this company do recognize themselves in the documentation of the Telecom Agency to be obliged to store traffic data for email services which they offer to their customers. But due to the technical design of their systems, there is in fact no data to retain.

> *'The formulation of the Telecom Agency on their website where they explain, it says that if you are a host provider you are obliged to log the mail data. [...]. One of the things that we've done is to change the server so that we don't log anything.'* – Provider

The researchers submitted the question to the Telecom Agency, whether webhosting would fall under retention laws. According to the Telecom Agency, this service is not covered by Section 13 of the Telecommunications Act. The Independent Postal and Telecommunications Authority doesn't record any webhosting companies in its database of public electronic communications services and/or networks either. However, according the Tele-

88 www.agentschaptelecom.nl/sites/default/files/eindrapport-nulmeting-wetbewaarplicht-telecommunicationsgegevens.pdf.

com Agency, this is different for email. A distinction is made between traditional email requiring an email application (e.g. Outlook) on any computer and webmail. The latter service is accessible via the Internet. The traditional email does fall within the scope of Section 13 of the Telecommunications Act as opposed to webmail, which doesn't. According to the Telecom Agency, there is no webmail service where signals are transmitted via electronic communications. Webmail is similar to a website with content. Websites also fall outside the scope of Chapter 13 Telecommunications Act (see also *Explanatory Memorandum (House of representatives)* 2006/07, 31 145, 3, p. 37). To date, the Telecom Agency has only encountered the webmail version of email at webhosting companies.

> *'The legislation has become "gray" with the new developments. A black and white answer is increasingly difficult to give. Developments in areas such as email go very fast. Nowadays you see all kinds of "hybrid" forms where traditional email and webmail are woven together. This is a trend that you see in a lot of services. The legislation is not adapted well enough to this.'* – Telecom Agency

4.3 Complexity of traffic and location data

During the interviews with the major providers, it became apparent that analyzing and interpreting the data requested was a task for specialists. Experts from both companies interviewed regularly act as telecom specialist or as an expert witness in litigation. The experts indicated that the complexity of traffic and location data sometimes causes problems with investigators due to their lack of knowledge and expertise.

> *'A research team collected data, which yielded conclusions [...] only when it comes before a judge do you get the real issues in focus. You don't know what all has gone wrong prior to that point. That's not to say that everybody is out there making mistakes on purpose, but the risk of error is simply huge.'* – Provider

> *'Some cases have not been tried, because mistakes had been made, information that had not been weighted, resulted in a very different decision.'* – Provider

> *'Call data is often misinterpreted. For example a customer calls in Rotterdam and suddenly you see a call in Groningen in between, because the system interprets the transmissions tower incorrectly.'* – Provider

The small provider, who was interviewed for this study, provides telephony services via the Internet. This is another form of telephony than the old-fashioned telephone services, and judging from the data requests received, it is apparent that the applicant's knowledge of this form of communication is not always sufficient.

> *'We received a selection number and when a subscriber dials this code before dialling his destination number, the call runs through us. We received a request for the incoming and outgoing traffic data, but a selection number doesn't run incoming traffic through us. Outgoing traffic could be retrieved via KPN or T-Mobile. So why submit the data request to us? It was a totally useless request. It doesn't cover the whole scope and half of it can't even be answered.'* – Provider

> *'Sometimes I have more information about a number, but the data request was not prepared properly.'* – Provider

The interviewee from the same small provider mentioned that he had recently received an international request for data, which his company purposely ignored. The requests contain the query or a foreign telephone number X, which was used to call via the Dutch network, and had contact with a number registered to that particular provider. According to the provider, the query of the data request was so general, that it could have been posed to any Dutch provider. The interviewee would have liked to have reported the incident to the officials, but did not know to whom to turn. This point was also raised by Data Protection Agency experts interviewed. There is a legal confidentiality regarding claims, but there is no legal provision that allows for the sharing of content information with supervisory organisations.
The incident described above was relayed to the National Interception Unit. Since the second half of the year 2010, all data requests go through the National Interception Unit, but apparently the system is not completely foolproof. According to an employee of the National Interception Unit, these types of broad and general data requests are filtered out. This information is reason for the National Interception Unit to reiterate among all parties involved that all claims for disclosure must be submitted via the National Interception Unit.

4.4 Irregularities

The functioning of the regulators and determining whether or not supervision has adequately been implemented, is beyond the scope of this study. No systematic study of all the major suppliers has been conducted nor that concerning the investigative services' requested output. While conducting

this study, the researchers did however, encounter some irregularities in the implementation of the law. For example, as per chance it was found that a large provider supplied traffic and location data including the *last cell ID*, i.e. the location at which a call is ended, despite this not having been requested in the disclosure claim, and there not being a necessary warrant. The law prescribes that only the *first cell*, or the starting position of a conversation, is covered by the Retention Act. Information about the location on which a call has ended can be very valuable to the police. Interviews with investigators showed that many found the last cell information to be valuable. However, information about the final location of a call is not covered by the data as described in the appendix to Article 13.2*a* Telecommunications Act and thus should not be supplied with requests for traffic and location data based on the Data Retention Act. This information can be requested under Article 13.4 paragraph 3 of the Telecommunications Act.
In addition, the researchers reported that three major providers supplied IP traffic data that were older than the retention period of six months. Though this is allowed in case of prepaid internet services, it is prohibited if there is a subscription to the services. The cases studied here, concerned advanced traffic of a smartphone with a subscription. Upon the request of historical traffic data for telephony the entire historical traffic IP data were also supplied. Apparently these providers didn't separate telephone and IP data in order to meet the different legal retention period requirements, possibly due to the extra work or to the technical complexity. In the meantime it has come to our knowledge that one of the large providers has adjusted the retention period for IP traffic.
The researchers have submitted these irregularities to the Telecom Agency, who were not aware of these issues. In practice, it appears to be difficult for the Telecom Agency to monitor compliance with the retention act. Under Article 18.7, paragraph Telecommunications Act, regulators have no authority request the traffic and location data from providers which is stored under Article 13.2*a* Telecommunications Act.[89] The Telecom Agency lacks an instrument to perform this aspect of supervision.

> *'For us it is very difficult to monitor this because we cannot retrieve that data. [...]. We can only control the process and that's what we do, but we are not authorized to access the real output of the process.'* – Telecom Agency

89 See also the explanatory memorandum of the telecommunications retention act (31 145, nr. 3, p. 55), as well as the explanation to Article 18.7, paragraph 2 in Knol & Zwenne (2013), Text and Commentary (*Tekst en Commentaar*) concerning telecommunications and privacy laws ('*Telecommunicatie en privacyrecht*') where the following point is addressed: 'In order to avoid any misunderstandings, and for the sake of the protection of the personal privacy of the electronic telecommunication service user, it is determined that the supervisory body has no authority with regard to these data.'

4.5 Private access to personal traffic and location data

Under the Privacy Protection Act, citizens are entitled to access the personal information that an organization stores about them. As can be read from the privacy statements of the various telecom agencies below, traffic and location data are accessible to private users.
Two researchers involved in this study, requested access to their own traffic and location data as mobile phone customers from their own telecom provider. A form to request access to personal information could be downloaded from the providers' website. After receiving their requests, both providers sent a complete list of all registered personal data. This included a complete list of contacts, type and number of the identification, contract and billing information, privacy settings on the identification number, IMEI and IMSI number and entry in directories. Furthermore, both providers supplied information about what data is stored under the Telecommunications Act and by whom they are retrievable. However, to gain access to actual traffic and location data, both researchers had to resubmit an application. Both providers were prepared to supply the requested data for a period of ten to fourteen days. An overview over a longer period of time would, according to the providers, comprise a disproportionate burden on the business. After sending a new request, no response was received form provider A. In light of the time frame of this study, it was decided to approach that provider via telephone eight weeks later. In that phone conversation the provider did not answer the question as to why no access was granted. Even after the customer had revealed himself to be a researcher of the Research and Documentation Centre, working on a report concerning retention laws, the call did not result in the requested information. However, five days after this phone call, the researcher received a letter containing the notice that it is not possible to give access citing Article 12.2*d* 'Processing of Personal Data of the General Conditions', which was enclosed with the letter. These terms and conditions concern the provider's use of the personal information for the following purposes: 'The analysis of the use of the network to the extent that the purpose of traffic management is necessary to ensure and improve the security and quality of service, and the continuity thereof, and the responsible business operation [of the provider].' An explanation as to why access to the requested information was refused that was understandable was not given. Provider B contacted his customer several days after submitting the request to access his data, asking what the reason was for the desired access. The researcher felt compelled to reveal his identity in the context of the study into the data retention act for the Research and Documentation Centre, after concluding that as a regular customer he wasn't going to get the information he requested. Within a week the provider sent the requested data. However, the names of third persons on the list of traffic data were made illegible to protect their privacy, somewhat strange considering that the numbers called

corresponded with names of people who the customer had obviously chosen to call himself. Additionally, the data provided showed on which date communication had taken place, the time of commencement of the communication and whether the communication was a telephone call, text message or traffic data. Based on the overview provided, it was not possible to determine location data. These columns were largely empty and no explanation of the cells that were filled in was provided.

Although the Privacy Protection Act is very clear about consumer rights to access such information, and the provider's responsibility to inform customers of these privacy policies, the providers approached by our researchers didn't deal with the customer data request appropriately. The researchers explained the situation to the Data Protection Agency, asking for their response to the situation that customers were getting little to no insight into their own traffic and location data. The Data Protection Agency was very clear, stating simply: 'That is unjustified.'[90]

> *'The legislative opinion emphasizes that it is important that the Minister recognizes that the right of access applies. In our view this was the only positive point of the legislative bill, that consumers at least would have the right to see what information about them is being stored.'* – Data Protection Agency

The Rathenau Institute has also expressed concern about the fact that legal right laid down by the Privacy Protection Act to access such information is in practice often no more than a paper reality. By the very refusal of access, it is impossible to check whether the processing of the data is accurate, complete, relevant and lawful. This affects the legal status of citizens and makes them largely dependent on the proper functioning of the system.[91]
An employee of Bits of Freedom (BoF) had a similar experience when requesting his own traffic and location data.[92] After submitting the request, this person was not given disclosure: only after pursuing a lawsuit the case was withdrawn by the provider, who then finally handed over the information requested.[93]

90 For more information on the Data Protections Agency's views concerning this consumer right see www.rejo.zenger.nl/focus/bemiddeling-door-het-cbp-inzage-persoonsgegevens-bij-telfort (consulted on July 26, 2013).

91 www.rathenau.nl/publicaties/publicatie/databases-over-ict-beloftes-informatiehonger-en-digitale-autonomie.html (consulted on July 25, 2013, also see www.rathenau.nl/uploads/tx_tferathenau/Hanf-out_ICT-commissie_Tweede_Kamer.pdf (consulted on July 3, 2013).

92 See also www.bof.nl/2013/03/15/drie-overgebleven-dataweigeraars-datadagboek-5 (consulted on June 12, 2013). In a later publication it became apparent that this person finally did obtain access to the requested data overview from his provider. See www.bof/2013/07/12/ohai-t-mobile-i-can-haz-my-data (consulted on July 15, 2013) and https://rejo.zenger.nl/focus/ik-de-ogen-van-t-mobile (consulted on December 5, 2012).

93 https://rejo.zenger.nl/focus/ik-de-ogen-van-t-mobile (consulted on December 5, 2012).

> *'What I find very important myself, is that the stakeholders, the citizens, aren't getting what they should be getting. [...]. The irony of the right to access as referred to in Article 35 of the Privacy Protection Act, and it's importance for the retention act, is that it does not work. This is because providers say it costs them too much effort to provide this information in a clear way to their customers. The only reason they do not have to give it is when it cannot reasonably be done, and it is precisely this which they rely upon. I think this is not correct. If investigative authorities request this information then the providers too have to provide the data easily, so I think the provider's processing operation is actually limited.'* – Bits of Freedom

A conversation with a provider yielded the comment that access to traffic and location data are actually unnecessary because the relevant data can be determined from a specified invoice. An interviewee from Bits of Freedom holds a different opinion:

> *'The Explanatory Memorandum states that the relevant data concerning calling behavior usually can be derived from the itemized invoice. That's nonsense because information on the invoice is only information concerning the amount of the bill, it does not provide information on received phone calls and messages nor does it provide information on location data.'* – Bits of Freedom

4.6 Conclusion

The data stored under the retention laws contains sensitive information about contacts and caller location of people. To ensure that data is properly stored, protected and destroyed, the providers have a legal obligation to provide technical and organizational security measures in order to prevent abuse of the data stored, and they are required to destroy the data after the retention period has expired. To prevent data falling into the wrong hands or data being used incorrectly, proper supervision of the implementation of such measure is important. The Data Protection Agency does not have an active policy concerning compliance with the obligations arising from the retention laws. Violation or reports are passed on to the Telecom Agency, which has been established in a covenant between The Data Protection Agency and the Telecom Agency. The Telecom Agency plays an active role as supervisor regarding compliance with retention laws and striving to make visitation inspections of all Dutch providers at least once every four years. However, the Telecom Agency only has the means to monitor the correct implementation of business processes, it does not have the tools necessary to ascertain a correct storage and provision in terms of the content of data. The

Telecom Agency does not have the authority to view the actual output of traffic and location data from different providers. The Telecom Agency lacks tools to be able to perform this aspect of supervision. When a government decides to store sensitive information of citizens, there must be a solid and effective supervision accompanying it. It is therefore recommended that the role of the regulator is strengthened in this regard.

Under the Privacy Protection Act, citizens are entitled to access an overview of the personal information that an organization holds about them. However, two requests for inspection of traffic and location data were barely honoured. According to the Data Protection Agency, this was unjustified: providers simply must honour requests submitted by customers. There is room for improvement concerning the handling by the providers of such requests.

A citizen should be able to obtain access to his own stored traffic and location data within a reasonable period of time, all the more because investigative authorities can access these data within a few days.

5 The use of historical traffic data in practice

In the following chapter the use of historical traffic data in the criminal investigation practice is described. The chapter is based on interviews conducted with key people on the deployment and use of historical traffic data. This chapter first discusses the use of historical telecommunications data in the criminal investigation process. It then discusses the use of historical internet data and transmission tower requests. Finally, this chapter will discuss the obligation of authorities to inform the persons in question of the fact that their historic traffic data have been requested, as well as the obligation to destroy data once the case is closed.

5.1 Historical telephony data

A summary of types of data that fall under the retention act is provided in the appendix to Article 13.2*a* of the Telecommunications Act. A strict distinction is made between telephony and internet data. For the sake of clarity, this dichotomy has been maintained throughout this report. However, in practice this distinction has almost disappeared and, according to experts, creates a false dichotomy in the Data Retention Act. In addition, telephony increasingly runs via the Internet. Previously people used Voice over IP services (VoIP) to make cheap calls to contacts abroad. But nowadays, many people subscribe to a cable company or provider whereby the phone traffic runs via the Internet. VoIP is thus a very common means of telephony. In the legislation bill where the retention period of internet data was reduced from twelve to six months, an exception was made for VoIP services,[94] for which the period of twelve months was maintained. This exception was made because the functionality of the telephony services can be regarded as a fixed or mobile network. The Experts Group Data Retention of the European Union (DatRet/expgrp. 2009)[95] recommended that the handling of VoIP services be harmonized for the different types of telephony. In effect this means that for traffic data retrieved from a phone that has called via the Internet, the same set of data is to be retained and supplied by the provider as when compared to telephony the 'old-fashioned' way. In the Netherlands the government distinguishes between different types of telephony by determining the functionality and the main features of the telephony service and prioritizing on the basis of this.[96] For telephony, such as traditional telephony, mobile telephony and VoIP, which for example uses the National Numbering Plan, making call forwarding possible, and offering access to the emergency service 112, a retention period of twelve months is maintained. For traffic from voice

94 *Explanatory Memorandum* (House of representatives), 2009/10, 32 185, nr. 3.
95 http://ec.europa/dgs/home/-affaires/what-we-do/policies/pdf/series_a_position_paper_5_final_14_07_2010_closer_understanding_of_the_term_internet_telephony_en.pdf (consulted on July 1, 2013).
96 See www.agentschaptelecom.nl/onderwerpen/veiligheid/opslag-telecomgegevens (consulted on July 1, 2013).

services using the internet and which do not have this functionality, a retention period of six months is maintained. This section will discuss the use and deployment of historical traffic on telecommunications specifically. Firstly, we will address the question of which data are stored.

5.1.1 *What is retained?*

In the appendix to Article 13.2*a* of the Telecommunications Act, a summary is provided of the data to be retained on telephone traffic. The law requires that the following information regarding telephony, should be stored by the service providers for a period of one year:

a telephone numbers of both the caller and the number that was called, and in the case of supplementary services such as call forwarding or call transfer, the number to which the call is routed;
b names and addresses of subscribers or registered users;
c date and time of commencement and end of the communication connection;
d type of phone service used (e.g. fixed, mobile or VoIP);
e in case of mobile telephony:
  - International Mobile Subscriber Identity (IMSI) and International Mobile Equipment Identity (IMEI) numbers (the identification numbers of the SIM card and the phone) of the caller and called party;
  - in case of pre-paid anonymous services, the date and time of the initial activation of the service and label (Cell ID) of the location from which the service was activated;
  - location label at the beginning of the connection (First Cell ID);
  - data for the identification of the location indicator.

The content of a conversation is not covered by the retention act, nor is the content of a text message, though the traffic data of the sent or received message is. A mobile phone is in regular contact with the network, even without it actively being used to make calls or to access the Internet. This intermediate class of data, called fleeting data, are not covered by the retention act and are thus not supplied for disclosure claims with the retrieval of historical traffic. Call attempts, where no contact has been established, are subject to the retention act and should thus be retained by the providers.

5.2 Telephony – overview of the use

The overall picture that emerges from the interviews is that historical telecommunications data play an important and highly valued role in the criminal investigation process. The interviewed professionals and experts provided many examples of situations in which such data can be used for investigative purposes. These stories exemplify how the data are used in very different ways during the investigation and prosecution of a diverse range of crimes.

Historical data on telecommunications traffic play an important role in the preparatory phase of investigations into organized crime. The data is used in the identification and mapping out of criminal organizations. Through a thorough analysis it is possible to chart who is in contact with whom. The locations from which calls are made also prove interesting because they aid in the mapping of individuals movements. During the preparation phase of a crime, suspects often do not realize that they can be the subject of a criminal investigation at a later stage. They are less wary. For this reason, accurate historical traffic data can sometimes still produce useful information, according to an interviewee. In offenses such as robbery traffic data are also used regularly for investigative purposes. For example, traffic data can be used to determine whether certain suspects were actually in the vicinity. In the case of a series of burglaries, location data can be requested on the basis of the location of the transmission mast. In this way, phone numbers from different crime scenes can be compared with each other in the hope that the suspects in several burglaries have been using the same phone. With such a tactic, an analysis is made of the phone numbers, or IMEI numbers, that appear more frequently in the provider's database, and subsequently a selection is made which numbers should be investigated further. This is a screening tactic that is used regularly in the investigation of serial crimes such as arson or a series of burglaries or robberies. This method is also used when there are two crime scenes that appear to be connected, for example when a getaway car is found at a different scene than the crime scene itself.

It may also be useful to request the historical traffic data of a stolen phone some time after a robbery. In this way it can be determined whether the stolen phone is still in use and whether the SIM card has been changed. There are examples of cases in which the use of this technique has led to a suspect. Historical traffic data can also determine whether a phone has been within a certain area, which in turn can be used to check witness statements for accuracy. In addition, historical traffic data can be used to check statements about who was last in contact with a victim or missing person. A police professional explained that in murder cases where suspects are thought to be related to the victim, the police often request traffic data from family and friends of the victim. By requesting these traffic data the whereabouts of such persons can be determined at the time the crime took place. This can yield an initial direction for the investigation, which can be leading in decisions made about who shall be invited for further questioning.

Following an arrest, a suspect's traffic data can be requested to determine whether or not the suspect has been in contact with another suspect. Persons who had frequent contact with a suspect, particularly around the time of the

crime, are of interest to the investigation and such information can be introduced tactically during the interrogation of a suspect.
It can also occur that during an investigation a witness states that he had been called by a certain person (at a specific time) who is wanted for questioning, but the witness does not have the number. In such a case, a public prosecutor can request to find out the telephone number on the basis of article 126*n* Criminal Procedural Code.
In addition to the examples mentioned above, the interviewees mentioned many more specific situations in which historical traffic data can play a useful role in the process of detecting and investigating crime. The list is certainly not complete nor is it exhaustive. These examples give an idea of the many ways in which historical telephony data can be used in the investigation process.

5.2.1 Considerations and goals

When an investigation team seeks access to historical telephone data, the application must be motivated. Amongst others, the motivation must specify the purpose of the traffic data requested. The investigation team must indicate which goal they hope to achieve with the data requested. These objectives can be classified into a number of broad categories, namely: identifying a user; finding out contacts, determining location, tracing an IMEI number, and making a capacity consideration before one proceeds with wire tapping. Obviously, claims for disclosure can serve more than one goal simultaneously.
One objective of the retrieval of historical user data is to *identify a user.* Often an investigation team only has a phone number, received by the Criminal Intelligence Unit, or a number that has been obtained from a phone tap of which it is unknown who uses it. Determining the identity of the user can be done in different ways. For example, the user registration of a phone number can be requested from the Telecommunications Research Information Centre (*Centraal Informatiepunt Onderzoek Telecommunicatie, CIOT*).
The Telecommunications Research Information Centre is the link between investigative services and telecom companies. Telecom and internet companies make lists of their customer base available (in a secure environment) to the Telecommunications Research Information Centre on a daily basis, including the statutory identification information. Identification data include names, addresses and residences associated with phone numbers, email addresses and IP addresses. The list of the customer base is kept by the telecom and internet companies for 24 hours in the secure environment and is automatically fed into the black box environment in the Telecommunications Research and Information Centre's information system. The information is refreshed every 24 hours, and re-fed into the Centre's information

system.[97] Identification data is attached to both fixed and mobile phone numbers with a subscription, and investigative authorities can request this information from the Centre.
The Telecommunications Research Information Centre has, however, two limitations. Firstly, the Centre only has current identifying data as the information system is refreshed and updated every 24 hours. Secondly, though prepaid phones are registered at the Telecommunications Research Information Centre, usually there are no identifying data coupled with them. It is possible for the Centre to ascertain to which provider the prepaid phone is issued.
Criminals often use (several) prepaid phones and SIM cards, which makes it difficult to identify the user of a particular phone. If user identification is not possible via the information systems of the Telecommunications Research Information Centre, it is possible to request the historical traffic data for a particular number from the provider via the National Interception Unit. Though having the National Interception Unit claim historical traffic data for a particular number will not yield any new identifying data, it will provide information for longer than just the 24 hours that are available to the Telecommunications Research Information Centre. The historical traffic data can also provide insight into the whereabouts of the phone, which in turn can assist in gaining more information about its user.

Historical traffic data can provide insight into the *contacts of a user* of a particular phone number. Historical traffic data shows the phone numbers with which contact has been made. Some of these contacts may be of interest to investigators, for example because they might be suspects or witnesses. It is also possible to gain insight into other still unknown contacts, which can be useful for the investigation. Often investigators are keen to gain an impression of the social network in which a suspect operates or in which a victim was active. Who he had contact with, when and with what frequency are important questions that can be answered. Sometimes a phone is only used for 'business' purposes, which allows investigators to gain a picture of an entire suspect network. On the other hand, sometimes telephones are only ever used to call a single other telephone number, from which investigators can deduce a secret one-on-one contact and/or suspect shielding strategies. In murder cases, or in the case of missing persons, victims' last contacts can be seen using historical traffic data, which can be useful to the investigation.
Another objective of historical data retrieval frequently mentioned by the interviewees in our study, is *determining location.* When a mobile phone conversation is started, contact is made with the transmission tower. This starting location, or first cell ID, falls under the retention act and is therefore retained by the provider. The transmission masts located throughout the

97 For more information on the Telecommunications Research Information Centre see www.rijksoverheid.nl/onderwerpen/telecomgegevens-voor-opsporing (consulted on July 1, 2013).

Netherlands each have a certain range which enables investigators to determine to a degree where the phone in question was located at the start of the conversation (see also section 4 of this chapter). Determining location through historical traffic data can provide answers to questions about where a user lives, whether a user was near the crime scene, what the users travel movements were, from what location a user called for the last time, et cetera. In murder or missing persons cases, knowing what the victims' location was at the time of the last call is potentially useful information to the investigative team.
Historical traffic data can also play a major role in considering a particular phone for wire tapping. Based on historical data on calling behaviour, an estimate can be made of the capacity required to listen to and analyze the tapped recordings. Additionally, traffic data can reveal that a particular phone is no longer in use. In that case, a tap on the respective number is superfluous. Finally, historical traffic data are often requested by the public prosecutor instead of a wire tap, when the crime isn't deemed serious enough to warrant a wire tap. In this way, the investigation team can still gain insight into the communication flows of a particular suspect.

5.2.2 *Which number to retrieve*

Historical traffic data can be retrieved using a phone number or IMEI number. Because suspects have sometimes had previous contacts with the police, and have thus previously given their phone numbers, the phone numbers of some suspects are sometimes already known and present in the police systems, such as Basic Investigation Provision (*Basisvoorziening Opsporing, BVO*), Basic Enforcement Provision (*Basisvoorziening Handhaving, BVH*), Blue View etcetera. It is also possible that a present suspect has filed a report of criminal activities in the past, and that his contact telephone number is thus stored in the police system. In addition, it is possible to request information on the basis of the unique identification (IMEI) number of the phone. Of course, it is also possible to get such information from the suspect himself, from a witness or the victim. In addition, telephone numbers can be obtained from tapped phone conversations or from historical traffic data of other phone numbers. In other words, a phone number can be mentioned during a telephone conversation that is being tapped, or a phone number can appear in the historical traffic data that is of interest to the police of which subsequently also traffic data can be requested. Another means of obtaining a phone number is via the Criminal Intelligence Unit (*Criminele Inlichtingen Eenheid, CIE*). This police unit has regular contact with the criminal world and thus regularly obtains information from there, including the phone numbers of certain persons. This information is passed on to an investigative team, who in turn can request the traffic data from the providers for those particular numbers. Finally, investigative teams can obtain phone numbers by analyzing seized telephones, determining whether or not there are numbers in them that warrant further investigation.

5.2.3 *Time of retrieval*

Historical traffic data can be retrieved throughout the investigation process, though generally the emphasis lies at the beginning of an investigation. Whether historical traffic is retrieved from the start of the investigation depends on how much information already is available at that time.
When not much information is available yet, but a phone number is available, it's relatively easy to map a social network using data from that phone number. That information can make the investigation move along, providing new lines of inquiry. In the case of a robbery where a telephone has been stolen, generally historical traffic data is retrieved quite quickly. When investigative teams make requests for historical traffic data early on in the investigation, usually not a lot of information is already available to guide the investigation.
Traffic data can also be obtained during the course of an investigation. Sometimes it turns out later that the traffic data appear to be a missing link in the investigation, and that they are thus duly requested. Coming across a new phone number by chance is also an occurrence leading to a traffic data request. One of the interviewees in our study explained that sometimes traffic data only get requested at the end of an investigation. She says:

> *'But, it can also only be done later, for example when the suspects have been detained and their phones have been confiscated. In that case, we look for IMEI numbers and other telephone numbers for which we request the historical data.'* – Police

5.2.4 *The principles of proportionality en subsidiarity*

If the investigation team wants to retrieve historical traffic data, it needs to obtain permission from the Prosecutor's office. To do this, the team's request must be sufficiently motivated. The retrieval of the data must be proportionate and subsidiary.[98] Basically, the investigation team is required to indicate the following: what kind of investigation it concerns, how it came in possession of the phone number, and what objective the team hopes to achieve with the historical traffic data. In general, if these questions are answered properly, the public prosecutor rarely rejects an application. The investigative officers interviewed for this study explained that because they never apply to retrieve information without reason, it follows that their requests are

98 The principle of proportionality dictates that the impact of investigative methods must be in proportion to the seriousness of the crime itself. The principle of subsidiarity refers to the question whether or not there is a different method to be used, which has less impact but achieves the same results.

almost always honoured. It does occur, however, that the public prosecutor telephones to request changes or amendments to an application, or that the time frame requested in the application needs to be shortened. Moreover, according to one interviewee in this study, the assessment of the application by the public prosecutor depends on his knowledge of the case at hand. Another interviewee said:

> *'If you're dealing with your own case officer [public prosecutor] who knows the case, then he generally just needs a quick reminder of what the case is about and it's done in a flash. But if I'm dealing with a new case that just came to light, and I call a random public prosecutor that just came on duty, then you often have to explain things quite extensively.'* – Police

5.2.5 *Frequency and age*

According to the respondents of our study, the retrieval of historical traffic data is done in each investigation where the proportionality and subsidiarity requirements are met. One of the professionals interviewed, working as an investigating officer, explained that the retrieval of historical traffic data is quite standard procedure. Other respondents say that a lot gets requested. This corresponds with the figures presented later in Chapter 6.

> *'I don't really have an idea about the frequency myself, other than to say that it is almost standard procedure.'* – Police

> *'Yes, we do that actually for each investigation. We use this type of information for every somewhat large investigation.'* – Police

> *'(...) not in every case, but in the majority of cases.'* – Lawyer

At the same time, respondents explain that information is not requested for every phone number that comes to their attention. In response to the question whether traffic data is requested for all telephone numbers in their possession and that is deemed important for the investigation, a professional replies the following:

> *'No, really, we are selective in that. I need to be sure that something is done with it. It really should have a purpose and not just be like, "We've confiscated three phones and let's request the data on them all." As you see often. The data overviews arrive and are not even opened. In the past that happened sometimes, but I am very selective in that.'* – Police

As part of the retention directive, it is interesting to determine how old the data that is being requested actually is. As can be seen in Table 1 of

Section 6.1.1, three quarters of the requested data is not older than six months. Respondents were also asked about the reasons for the period for which information was requested. This seems to tie in closely with the nature of the investigation and the nature of the research question. According to respondents from the police department, when investigating a robbery, generally standard data is requested covering a very short period of time. An investigative professional gave an example of a street robbery in which a mobile phone was stolen. In this case, the historical data from the stolen phone is requested for several days to try to determine the location of the phone and who is using it. Other interviewees stated that they retrieve standard data over a period from several months up to six months. This is generally the case for investigations that last longer, such as in murder cases and investigations into organized crime. In response to the question 'How far back in time do you request information?' an investigative officer replied:

> *'That depends on your investigation. You always make up the balance. If I have a very specific question, I have a murder case. And if I want to know something about the contacts of the week before, I just request information for that week. If you're trying to analyze who someone is, you want to identify them, who their contacts are, then you retrieve information over a longer period. Sometimes you want to map contacts, let's say someone says that they call each other once a month. Well, if it is important to us, then we request data for a year and you can see then if it turns out that someone calls ten times a month for the whole year, which gives a very different picture. So it really depends on what your question is what you request. It is never standard like "Oh let's request one year". That just isn't always helpful.'* – Police

5.2.6 Data analysis

Investigation teams receive historical traffic data digitally. The National Interception Unit is merely a middleman, where requested and supplied data come together and get sent on to their respective parties. Decryption takes place here, but no analysis of the files. Traffic data is not always provided in the same format. It can take a lot of effort to get everything into the same format that is necessary to further analyze and interpret the data. Some providers send the encrypted traffic through to the National Interception Unit, from which the files will be forwarded to the police via a closed secure network. In general, the requested information is encrypted and sent securely. However, according to the National Interception Unit expert, some providers do not use secure transmission as directed by the Interception Unit. In those cases, unencrypted traffic data is sent via the email.
There is no fixed standard format in which data is sent. Each provider supplies requested data in his own way. According to some experts in the field,

analyzing the files is sometimes difficult. Files can be analyzed either manually or with the aid of an analysis program. Most respondents indicated that they use Digital Communications Traces, an analysis software program, to analyze the traffic data. The incoming traffic is considered specialist work that is usually done by an analyst. However, not in all investigations an analyst is involved, so sometimes the traffic data is actually analyzed by investigation officers.
Once the files have been imported into the Digital Communications Trace program, the person working with it can formulate the questions that the investigation team wants answers to, such as: which contacts are associated with this number, which transmission towers have connected with this number and so on. The finished analysis using the Digital Communications Trace program cannot be used as evidence. To do that, one needs to return to the original format (if possible)[99] which can serve as proof. One respondent said about this:

> *'I think the use of Digital Communications Trace has also been accepted and approved by the judiciary and yet it is still said that you have to go back to the original. I can understand that, because the Digital Communications Trace makes an interpretation, it performs an extra step. I can also make an estimate, the Digital Communications Trace can't do anything strange with that. It's a call from A to B, at a quick glance to the original I am convinced. If it is really important about number of text messages, I have to sift through all the original messages. When it comes from [provider X], I won't even bother. That is supplied in xml-format, I can't show you now, but it's quite depressing. (...).'* – Police

Professionals and experts from the police find that going back to the original document is very laborious and, according to one of the investigation officers, it boils down to a duplication of work. Considering the difficulty that specialists experience with this task, it begs the question if lawyers, public prosecutors and judges are actually able to assess the value of original traffic data. This question was put to lawyers.

> *'In a plea of a case I see that the expert spoke of the "sparse nature of the Telfort network; relationships with other base stations; about position point 15,913; the nearer base station 48,753, and so on". Well, you get to work on that, but it drives you crazy'* – Lawyer

Police experts regard analyzing the historical traffic data as specialized work, only to be done by police analysts. The police academy offers a special training for working with historical traffic data, where officers learn to use the

99 Some providers supply traffic data in xml-format. This is very difficult to interpret without a computer software program.

Digital Communications Trace software. Due to rapid changes in the phone market, regular updates of the course are offered. According to respondents, apart from the training, it is also a question of delving into the subject and exchanging information with colleagues, to gain a deeper understanding of how it works. The two lawyers we interviewed for this study both had had experience with experts in the field of traffic and location data. However, due to the high costs involved, most of their clients were not able to afford such expert witnesses. One lawyer says he decided to map the information himself.

> *'If it is in the interest of the case, and the defendant is adamant about it being wrong, I don't ignore it. We had a murder case a few years ago and I mapped the connections again myself, drawing lines on the basis of the expert witness testimonies given to the magistrate. It turned out that the area where my client actually had found himself was actually quite different. It was a much smaller area because; apparently a transmission tower had not been functioning at that time. [...]. So then you realize that the hard evidence isn't really as hard as it had appeared to be. Often it is presented as indisputable technical evidence, as objective evidence, and though in principle that may be so, the reality is that the interpretation thereof is very complicated.'* – Lawyer

> *'You can ask an expert. They are able to analyze the raw data, but experts are very expensive, easily costing a few thousand Euros. The client must be able to pay that. Sometimes they can, but often not. Usually the money just isn't there.'* – Lawyer

It is also possible to check the analyses and interpretations in the police reports. Generally it is the analysis and interpretation that is questioned.

> *'The risk is that you get a biased presentation of such data. You can ask to check the data, but then you get the raw data, which boils down to hundreds of pages completely undecipherable transmission data for which you need software to analyze sensibly. As a defence attorney you don't have access to that, so in effect you can't do anything with the data, while the prosecution maintains that you have the opportunity to check it. There lies the risk, in the instrument as well as in the way that it is interpreted. [...]. Theoretically, as a lawyer, you should be given the opportunity to suggest what rolls out of the analyses if you throw in your own hypothesis. But that isn't possible.'* – Lawyer

> *'Generally you receive an overview of the police report that states, "We have conducted a telecommunications research and conclude the following." If these conclusions are disputed by the client, we want to see the underlying*

material and check the information. [...]. The chances of scoring with that are not so great in my experience, as the information is generally correct. In fact it is generally the subjective conclusion drawn from the data that is questionable.' – Lawyer

5.2.7 The revenues

Historical traffic data for telephony are considered to be very valuable by all interviewed professionals and experts. Depending on the phase of the investigation, it provides information that can guide the investigation further along or it serves as evidence. At the beginning of an investigation, historical traffic data provide useful information upon which further investigation can be initiated. On the basis of such information, it can become clear who are possible suspects and where the observation team should stake out. A professional of the police department says the following:

'It really can be an advantage. It is never the only evidence that you have, because that would not be good, but it is a valuable support in the investigation, in contact between phone numbers and possibly between people. It can serve as an aid in selection. It can save you a lot of work. [...]. If you do not know where someone lives, then you look at the first and last connection of the day and then you look at the transmission tower and if I see a line in there, than I know approximately where someone goes to bed. I can map that information and thus send the observation team to this place instead of that. That makes it efficient. It helps you focus your investigation just a bit more at times.' – Police

When the investigation is in a more advanced stage, the historical traffic data also serve as evidence. The interviewed experts and professionals speak particularly about their supporting value as evidence. Historical traffic data is never the only evidence, it merely serves to strengthen the case already there. A good example of evidence provided by historical traffic data, is given by the police professional below. She tells about an investigation into phishing:

'The bank customer received an email with a cloned website, where he was supposed to fill his information. He was then called by a so-called bank employee. It went something like this: "I'm Mrs. Smith of ABN AMRO and I work in security matters. We are currently in the process of updating everything for security purposes. Would you please get your bank card and your identifier. Then we can proceed to walk through this." People are still falling for this. We found a lot of SIM cards and telephones at the house of one of the suspects we were investigating. Upon analyzing the requested historical traffic data for those phones, it became apparent that many victims

had been called with these phones. You can add that nicely to your case file and also for the public prosecutor.' – Police

In accordance with what is mentioned in the section earlier concerning considerations and goals, the interviewed experts and professionals express that the revenues of historical traffic data are precisely those points: (1) identifying a user, (2) tracing connections/contacts; (3) location determination, (4) tracing an IMEI number, and (5) conducting a capacity investigation before proceeding to wire tap. It does sometimes occur that historical traffic data is requested but in the end is not viewed or analyzed. That is inherent to police work, according to one respondent:

'It doesn't happen often, but I have experienced it. Sometimes you really intend to use it, for example, to identify or locate someone. Only it's not the only lead you've got. You also have an observation team, for example, or other investigative means, so sometimes you find out someone's identity or whereabouts before you get around to using the historical traffic data. That happens, but you can't presume that in advance. Sometimes one way is faster than the other.' – Police

Other professionals state that this does not happen to them. If data is requested, it is always viewed. One interviewee said that it was something that occurred in the past, but that that is no longer the case. A detective must properly justify what he will do with the traffic data, and say what he intends to achieve with it.

5.2.8 Relevance of retained data

The retention act requires providers to retain and store certain data to make it available for a set period for investigative purposes. We asked our experts and professionals whether they considered all the stored data to be equally relevant or if they felt that certain important information was missing.
All of our respondents indicated that they considered the stored telephony traffic data to be highly relevant. An investigative professional, however, stated that he never used the X and Y coordinates of the transmission towers. According to the respondent, the national unit has a database, dubbed Route 66, in which the locations of all the transmission towers are marked, making the X and Y coordinates redundant. A number of interviewees indicated only knowing whether the stored traffic data are relevant upon viewing it. Sometimes, upon receiving the traffic data for a certain phone number, it turns out to be useless, because, for example, the phone isn't in use.

A number of criminal investigation professionals indicated not only being interested in the starting location (first cell) of a telephone conversation, but also wanting to receive the final location (last cell). The starting location, which is the transmission tower with which is connected at the beginning of a conversation, is covered by the retention act. However, where the conversation ends, or the last connection to a network base, is, according to the appendix to Article 13.2*a* of the Telecommunications Act, not covered by the retention act. This means that when, whilst making a telephone call, someone gets into a car or boards a train, it will not be possible to determine the end location of that caller on the basis of the historical traffic data. However, the final location of a call is apparently included in the data retrieved and supplied by one of the major providers.[100] This is a noteworthy point, as a conversation with an employee of the Telecom Agency made clear that in the past this provider had been reprimanded for this error and had been forbidden to include last cell (final location data) with the supplied traffic data. However, the final location of a call can be requested using a different disclosure claim.

5.2.9 More efficient investigation?

The retrieval of historical traffic data is one of the many means of investigation used by the police. When asked whether the retrieval of historical traffic data contributes to a more efficient investigation process, most experts and professionals interviewed responded affirmatively. The revenues from historical traffic data enable other investigative procedures to be refined and better targeted. When the traffic data allow the investigative observation team to refine a search to an area, as opposed to an entire city, it makes it more efficient. If it becomes apparent that a telephone is hardly used, there is no need for a wiretap to be requested. And locating a stolen cell phone greatly assists in tracking down the suspects.
The retention act has facilitated a more efficient investigation, according to a number of professionals and experts, because the telephony traffic data are now retained and available to be requested for a year. Before the law went into effect, determining both the retention periods and the destruction of traffic data was left up to the providers themselves. There had been a limited retention of three months for prepaid numbers, however, but the respondents indicated that before the retention act traffic data were generally available for a shorter period of time than they are today. A professional says the following:

> *'[...] Three months go by in a flash. Even for investigations very much focussed on the present, and also because the legal profession might make request for something. This is also a good point to emphasize. Lawyers can say that their client never had contact. And if those data have been destroyed, they no longer exist, the year is over, you can say what you want,*

100 See also Chapter 4, section 4.1.

but it cannot be confirmed or refuted. And the defence team only comes into action later on in the game. In that regard, I can imagine that it is a good thing that the retention period is one year.' – Public prosecutor

'Now I'm sure that if I request something, I can go back a year. That used to be a bit uncertain. That was more chaotic, then you were really dependent on the provider. Now it's plain and clear.' – Police

5.2.10 Is the retention period sufficient for investation in telephony?

The European Directive leaves Member States the choice in retention period ranging from six to twenty-four months. The Netherlands has opted for a retention period of 12 months for telephony data.
Both during the interviews and on the basis of the statistics presented in Chapter 6, it can be deduced that investigative services sometimes have a need for traffic data from a time that postdates the expiry of current retention periods. As can be seen in Table 1 of Section 6.1.1, a number of data requests have been made after expiration of the retention period. This doesn't violate any laws or regulations. One may submit a claim further back than the retention period covers. The need for these older traffic data occurs particularly for cold cases, in which investigations have been referred back by the Court of Appeal who may request new or additional research to be conducted, or for cases in which after a year suddenly a tip is received. The interviews revealed that the majority of police professionals believe the retention period of one year to be sufficient for the work that they do. They do however, point out that though the retention period is long, there are cases where for longer-term investigations it is insufficient.
One interviewee gave the example that a whole year of a suspect's traffic data had been requested in an investigation into a robbery. Despite this, the retention period of a year was not enough:

'In that investigation we would have preferred to go back a bit longer, because as it appears, the car which was used for the robbery, had also been used for a robbery that took place just prior to the start of that year. We only found that out recently and regard this as a missed opportunity. But anyway, how long should go on storing data.' – Police

However, some investigative teams, such as a street robbery investigation team, never request the historical traffic data for a longer period and thus indicate that for their purposes the retention period may even be shortened. The general consensus between investigative experts and professionals is that the retention period of one year for telephony traffic and location data is sufficient and shouldn't be shorter.

5.2.11 Notification and destroying data

Because special investigative powers pose a potential breach of privacy, the public prosecutor is required by law to notify persons under investigation that such authorities are being exercised in their name (Article 126*bb*, section 1 CCP). Once the case has been closed, all information obtained under the authority of Article 126*cc* paragraph 2 of the Code of Criminal Procedure must be destroyed. These laws also apply to the disclosure claim for historical traffic data. The authority to recover user data is exempt from an obligation to notify such persons (Article 126*na/ua/zi* CCP).

The notification method has not been explicitly examined in this study. An extensive study on phone and internet tapping published in 2012, provides a detailed description concerning notification (Odinot et al., 2012, p. 138). An important finding was that, though generally the office of the public prosecutor does comply with the notification obligation, it doesn't pose a high priority.

According to the interviewees, even in cases that do not end up going to trial and in serious crime cases, in principle persons are notified, 'provided that the investigation is not jeopardised by it'. A respondent working at the 'special investigative powers office' and dealing with notifications on a full time basis, explained that only in about ten to twenty cases per year, out of about 2,000 a year in that particular area, his office decides not to notify persons. According to the same respondent, notification didn't occur for any cold cases, for cases of embargo and for sexual abuse cases. Obviously, failure to notify can also be the result of an inability for the authorities to track down the necessary addresses.

The police officers and public prosecutors who were interviewed for this study were asked about the obligation to destroy data three months after notification. They responded stating that all data obtained under the authority of these laws, such as requested historical traffic data, are indeed ultimately destroyed. However, the destruction does not always go smoothly: certainly in the case of old investigations it proved difficult to unearth all the information. An expert from the police explained that this is primarily due to a technical problem, that ICT systems are not installed for this. In his opinion, such data should be given a digital label, such that the software system can recognise it and it can be destroyed accordingly. Historical traffic data are often put in police systems, the data would have to be removed from these systems as well. The interviewed expert notes that this does not always happen, simply because the information is hard to find. Marking the information by attaching a digital label would help.

5.3 The use of historical internet traffic data

Besides telecommunications traffic data, investigation services can also retrieve historical traffic data on internet communications. This section discusses the way in which historical data on internet traffic is used in the criminal investigation practice. We will address the question of precisely which data are stored and retrieved. Additionally, we will provide an outline of the way in which historical internet traffic data is used in the investigation and prosecution of various types of crimes.

5.3.1 What is retained?

The Telecommunications Act mandates that data relating to internet access, email via the internet and internet telephony be retained by the providers for a period of six months. As noted earlier, it appears that at some points the law has been overtaken by technological developments and by changes in telephone and internet use. This is particularly true for the part of the retention relating to internet traffic data. In the appendix to Article 13.2*a* of the Telecommunications Act for example, internet telephony is listed under the section on internet traffic data. In the explanatory memorandum to the legislative bill an exception in the retention period of internet data is made for VoIP services, of which the functionality so closely resembles that of traditional telephony, such that these services can be regarded as telephony over a fixed or mobile network. For these services, a retention period of twelve months remained valid.[101] As a result of this decision, and on the recommendation of the Data Retention Experts Group of the European Union, Dutch providers agreed to harmonise the definitions of the different forms of telephony (DatRet/expgrp 2009).[102] In practice this means that when traffic data is requested from a phone that has called via the internet using a service that meets certain functionality requirements, a similar set of data is provided as compared to a disclosure claim for 'old-fashioned' telephony. This has been made possible by making certain distinctions in telephony, defined by the government. These distinctions are based on the functionality of various forms of telephony and the main features of the telephony services.[103] For VoIP services which, for example, use the National Numbering Plan, which allows call forwarding and offers access to the emergency number 112, a retention period of twelve months is maintained. For traffic data from voice services that use the internet but do not have this functionality, a retention period of six months is maintained. There are also internet telephony services such as Skype, Icall, Axvoice, FaceTime, etc. that are excluded from

101 *Explanatory Memorandum* (House of representatives), 2009/10, 32 185, nr. 3.
102 http://ec.europa.eu/dgs/home-affairs/what-we-do/policies/pdf/series_a_position_paper_5_final_14_07_2010_closer_understanding_of _the_term_internet_telephony_en.pdf (consulted on February 1, 2013).
103 See: www.agentschaptelecom.nl/onderwerpen/veiligheid/opslag-telecomgegevens (consulted on March 4, 2013).

the retention directive because the services are not subject to retention laws or because the provider is located abroad.
As described in Article13.2*a*, appendix B of the Telecommunications Act, the following data is stored for a period of six months:

a assigned user ID (IP address) and the user ID or telephone number of the intended recipient of an internet telephony call;
b the user ID and telephone number allocated to any communication entering the public telephone network;
c the name and address of the subscriber or registered user to whom the IP address, the user ID or the telephone number is allocated at the time of communication, and the name and address of the subscriber, or registered user, and the user ID of the intended recipient of communications;
d the date and time of the log-in and log-off of the internet sessions based on a certain time zone, together with the IP address, either static or dynamic, assigned by the provider of internet access service in a communication and the user ID of the subscriber or registered user;
e the date and time of log-in and log-off to an email service or internet telephony service based on a certain time zone;
f the internet service used (Wi-Fi, dial-up, mobile, etc.);
g the in-calling phone number for dial-up access;
h the digital subscriber line (DSL) or another end point of the originator of the communication.

Historical traffic data on internet and email use can provide insight into the IP addresses used by a particular person, as well as into the email contacts of the transmitter and the receiver. The content of calls, messages or emails, search terms typed into search engines, and IP addresses of visited web sites are not subject to retention.

5.3.2 *Relatively little use of internet traffic data*

During the interviews conducted for this study, it became clear that the interviewed professionals had extensive experience with working with historical telephony data, but had little to no knowledge of how historical internet data could be used in criminal investigation processes.

> *'It is unknown to me that IP data also fall under the retention directive and I'd have to look up what information you would get. Where do I get that? Do I get it just like that? Do I get it in my other tap system? I have no idea.'* – Police

> *'It strikes me as very interesting. Only I have no idea of what I would get.'* – Police

> *'It's hardly used; it is something of recent times. This is hardly used in my practice.'* – Lawyer.

Interviews with those familiar with internet data retention and the way in which it can be used in the criminal investigation process, revealed that even those investigators preferred to use historical phone data as opposed to internet data. This is also reflected in the figures for 2012 (see Section 6.1.1, Table 1), showing that the number of requests for internet and email traffic is much lower than the number of requests for telephony data.

> *'Data from IP is used sometimes, but if you compare it to the number of telephony requests, it's really just a fraction, though it is important. [...]. In particular, it's important if there has been certain activity from an IP address.'* – Police

> *'Yes, I do sometimes use IP data, but not as often as telecom data.'* – Police

> 'We do it sometimes. It [requesting IP information] isn't very common, but we do do it.' – Police

> *'(...). All investigators also have mobile phones and laptops, but it is still very classic, mainly telephony that is being looked at.'* – Public prosecutor

Many professionals seem to be apprehensive and have a lack of knowledge about how internet traffic data can be used in the criminal investigation process. This is confirmed by one of the public prosecutors interviewed.

> *'Not only do you need a lot of knowledge about what the possibilities are, but also about the routes that you should follow to get the information you need. I notice that in the field sometimes and it seems to be very persistent. [...]. I think that a large part of the detectives don't have that knowledge, and that they don't know that they can pursue further [investigate further].'* – Public prosecutor

> *'We've invested a lot, as have the National Interception Unit and the regional interception coordinators, to boost the level of knowledge, to make clear to people what to do and how. But for some reason, it's very persistent. I don't know exactly why, frankly.'* – Public prosecutor

This did not come as a surprise to us. Similar findings have been published in 2012 in a report on the use of telephone and internet tapping in criminal investigations (Odinot et al., 2012). That study showed that few investigators are familiar with internet tapping and that criminal investigation on the internet is still in its infancy. Moreover, activities related to internet issues are often performed by experts because the digitization of today's society has not yet become part of the daily scope of many investigators. There is still much to be gained on this matter, for example by hiring young people who are part of the digital society, but also by targeted training and other forms of transference of knowledge.

> *'The complaint that investigators nowadays are not able to do much more because a lot goes via internet, is nothing in comparison to the amount of extra data that they have already received and which they can access much easier. Due to the same technology, a lot has also become much easier. I think it would be good for the police to pay more attention to their own knowledge level.'* – Bits of Freedom

At the same time, we must take into account that technology develops at a rapid pace. So fast in fact, that even for the few experts themselves it is difficult to keep up. This makes it difficult to bridge the investigators' knowledge gap.

> *'There really is a glaring shortage of digital experts. It is also difficult for us to keep up. This year I will again attend an internet training course but we can't keep assisting all the detectives for all their cases. So we remain with a serious backlog.'* – Police

> *'There are few people who have sufficient knowledge. Fortunately, we also have a digital expertise department. They are the people specialized in computers and reading the hardware. Luckily, they have a working knowledge of network traffic, upon which we also sometimes rely. But even still, there are too few people to get the most out of it.'* – Police

> *'You mustn't be thinking in your own world, for the generation behind us, the 14 and 15 year olds have no money, but they always have internet and know exactly where to find free ways to log on.'* – Provider

5.3.3 Considerations and goals

Historical internet data can be obtained from a provider via the National Interception Unit. Historical internet traffic data are often requested for crimes and offenses committed via or using the Internet. This can cover such topics as sending threatening emails, internet scams, human trafficking or distributing child pornography. Deciding whether or not to start an internet tap however, may also be a reason to request internet or email traffic data. This is to determine in advance whether the internet address or mailbox in question is actually in use.

Generally, identifying a user or a connection is the main reason for requesting information, according to the experts and professionals. When an internet address is visited, sometimes investigators want to know who is behind a particular IP address. An example provided by one of the respondents in this study, is of a case of fraud with childcare fees. In this case, the IP address of a particular visitor to the tax authorities site was traced. By identifying the user of this IP address, the authorities were able to identify a suspect and make an arrest. For cases of internet fraud internet traffic data is regularly requested, whereby the identification of the user and the time at which a website was visited, is considered valuable information.
Another expert refers to a human trafficking case, whereby a home computer was used for criminal activities. A further example can be seen in a major child abuse case, whereby the suspects used techniques to enable them to maintain anonymous contact with other paedophiles. Also in this case, internet data was used successfully to identify a suspect, despite shielding techniques used by the suspect. This was due to human error on the part of the suspect, who forgot to use the service that concealed his identity.
Usually, visiting a particular website by a certain IP-address can be shown by the use of a combination of data, such as date and time information, sometimes supplemented with an email address. The combination of data requires a smart and thoughtful analysis of the stored data so that it can serve as supporting evidence. After the IP address has been located, it is up to investigative officers to locate the accused using this address. Of course, it is possible that these people don't actually live at the address associated with their IP, for example, when users login through that Wi-Fi address or to use the IP address as a guest. It is possible to do this without the knowledge of the owner of an IP address. One of the respondents noted a case where an IP address had been logged onto without knowledge or consent of its owner, due to the poor security of the owner's network.
The investigations described above, all of which were aimed at identifying a suspect behind an IP address, have one thing in common. In all cases, the purpose was to track user and traffic data from a *fixed* IP address. These IP addresses are usually the same for a long time and additionally, for a home internet connection, registration and subscription to a provider are required. In contrast to mobile internet, the terminal user of a fixed IP address is known and easily traceable through the service provider or via the Telecommunications Research Information Centre.

5.3.4 Mobile internet

The interviews revealed that identifying a user of mobile internet through historical traffic data is difficult and often even impossible. Unlike a fixed internet connection, where IP addresses are associated with a user for a longer period of time, the IP addresses of mobile internet users are dynamic. This means that internet addresses change frequently. In addition, due to a shortage of IPv4 addresses, often multiple users are grouped under one IP address, which makes it impossible to identify the individual user. Moreover, many mobile internet users regularly use Wi-Fi networks or hotspots. These networks are offered in more and more places, and they are often free. Hotspots are not always public; they are often only available to customers of the provider of the hotspot. Wi-Fi networks that are not public, aren't covered by the data retention act. However, when a device is logged on to a Wi-Fi network, it communicates via the internet using the same IP address as all other users of that network. It is thus not possible to identify an individual solely on the basis of an IP address derived from internet traffic data.

> *'For dynamic IP addresses, it is obviously very difficult, especially when hotspots are frequently used for mobile internet.'* – Public prosecutor

> *'Yes, we have problems with mobile internet. It does not make it easier. In some cases we succeed, but definitely not always.'* – Police

> *'You can't come to know with Wi-Fi, and the phones themselves use variable IP addresses. So that doesn't help very much. Usually, a connection is made with the provider and the provider then connects to the Internet. For example, you get an IP address from KPN on location in The Hague or so. So that really isn't very helpful.'* – Police

However, according to one expert, when the traffic data reveals an IP address indicating a hotspot, such as 'KPN Hotspot Co.', it doesn't mean one can't look further.

> *'We will know that we have no direct link, but then we also know that we need to request information from the provider like, "Can you tell us which IP addresses or MAC addresses had contact with that hotspot in a that particular period?" [...]. Detectives don't know that they can follow through in their information requests with more questions. So they stop after the first request, but then don't know that there is also a second, and that all they have to do is make a phone call to get some customized answers. Because that's what it is, a customized request.'* – Public prosecutor

It should be noted that the data of individual users logged in to a hot spot are not covered by the retention act.
The police interviewed indicated that historical internet traffic data for mobile internet also is requested to determine the location of a particular smartphone. When a smartphone makes a connection with a transmission tower, in the traffic data is registered with which tower a connect was made. Thus, the location of the device can be determined fairly accurately from the

traffic data. Even when someone does not use the phone to call but only to communicate via the internet, location data is generated when contact is made with a transmission tower. However, due to underlying technical reasons, these data are considered to be less reliable than telecommunications location data.

> *'A "histo" from an internet session using a smartphone is quite useless content wise, because you only see that there was an IP session between a phone and a provider, but you don't see what happens behind that. But details about the location, which can be very important in a criminal investigation, where someone was located, is included in the "histo".'*
> – Police

It is also possible to use internet via a prepaid mobile connection. In that case, the provider has no information about the user whatsoever. The most pressing question for investigative purposes, namely identifying a user, seems to be impossible to answer in this case. Having said that, for a seized telephone with prepaid internet, or a telephone found on a victim, it is of course possible to retrieve the traffic data from that phone.

5.3.5 Email

Information on the sender and the receiver of email traffic is also kept under the retention act. However, none of the respondents in this study indicated to ever make use of this information to find out who has been in contact with whom. The only time this type of information is requested is to determine if a mailbox is actually being used. If it appears that the mailbox is being used, our respondents indicated requesting a 'mailbox dump' from the Dutch provider, in order to recover the email traffic data. This includes the contents of the mail that doesn't fall under the retention act.

> *'On occasion, a team looking into the Telecommunications Research Information Centre will notice an email address as a customer at Planet, for example. [...]. The only thing that you want to know is if this mailbox is actively being used, or being accessed. A yes or no question from the provider would be sufficient to determine whether or not to tap that address. In practice, however, it doesn't get that far, because it appears not to be in use.'* – Police

> *'In my experience, if a suspect is younger than 55 years old, chances are he won't be using that Dutch email account.'* – Police

> *'When impounding a computer we have full mailboxes, going back many years at that. In that regard, email traffic data is of little use to us, because*

in the headers and the text, we have access to where the message is going and who it's from.' – Fiscal Intelligence and Investigations Services

'When you're dealing with a victim and you want to know with whom he/she has been in contact, and impound their email from where they always use it at home [using a Dutch provider], then that is important information. But these aren't the criminals of course. The criminals know what not to do. But victims, innocent citizens. They leave so many personal traces behind in their email, it's huge.' – Provider

5.3.6 The usability of retained data

Whether one makes use of historical internet traffic data depends on the nature of the crime. When an offence committed has something to do with internet, internet traffic data is retrieved. This seems logical, but telephony traffic data are used for a diverse range of crimes, and not only when an offense has something to do with telephony. And this is so, despite the fact that a major part of communications are conducted via the internet. According to several experts, this is due to the fact that the internet traffic data retained is only partially suitable for criminal investigation purposes. According to the respondents, most of the information, as described in the appendix to Article 13.2*a* of the Telecommunications Act, to be retained pertaining to internet use, is outdated. The directives dating from 2009 don't match the current use of the internet, nor do they match technological developments in the field. An example hereof can be seen in the log-in and log-off data for internet and email sessions. In the past, when a modem needed to be actively switched on in order to connect to the internet, this information was relevant. Nowadays, however, now most people are connected to the internet 24 hours a day, 7 days a week via Wi-Fi networks, for example, and there is only talk of logging in or off when the device is turned on or off. With the changing range of internet services, and the subsequent changes in internet behaviour, the value of the retention directive pertaining to internet data for criminal investigation purposes, has decreased.

'The data retention directive dates to the beginning of this decade and actually looks very much to the situation of the preceding period; so dial-up connections, pop-mail, and short sessions of connection in order to pick up emails. Email was also very important then. So the data that is stored these days, a decade later, is actually more relevant to the circumstances of the late 90's, early 2000's, and which information was considered important for detection back then.' – Police

'The main focus of the retention directive pertained to the communications traffic data which we require providers to retain. And because no one could

imagine then that we would all be using social media ... no one thought about that then. Email was new, and people still used to use the phone a lot.' – Police

'Text messaging is almost dead among our youth. Calling by phone is also decreasing according to the Independent Postal and Telecommunications authorities. It isn't strange that 'Hi' is issuing stickers these days with the slogan 'Who calls any more these days?' Obviously they do that to promote internet subscriptions. These days you only ever see people using What's App ping or Facebook. KPN mail is rarely used anymore. Communications services have shifted to providers on the Internet. It's all web-based.'
– Dutch Forensic Institute

Dutch Internet Service Providers, as defined in the Telecommunications Act, must adhere to the retention directives pertaining to traffic data, such that the data can be made available to investigative services for the purpose of assisting in the detection of crime. However, when persons use common foreign Internet Service Providers like Gmail, Hotmail, Facebook, and the like, no traffic data is available, because the companies do not fall under Dutch jurisdiction. These international providers hold a prominent position in today's internet landscape, of which the Dutch providers only hold a modest place. It is noteworthy to mention that several foreign Internet Service Providers actively and explicitly focus on the Dutch consumer market, holding large interests here. They are able to avoid retention obligations by hiding behind their foreign parent company. This holds true for tapping obligations as well. The Dutch retention act is limited in its jurisdiction by borders, whereas the internet isn't.

'Google Netherlands simply says "No, we don't have those [traffic] data. You'll have to get it from the United States." That of course, seems a bit odd.' – Police

'[...]. Because technical developments move so fast, the law is simply not flexible enough to deal with it and so we miss a lot of the information. [...]. What I really want to say is: chapter 13 should be adjusted and then the concept of provider be changed.' – Police

'[...]. It's all web based now. I don't understand why politicians, and then specifically those in Brussels, just let them get away with this. There, where the online providers are, a lot of money is being earned. For advertising and marketing all kinds of user information is being used, but when we ask for even just a fraction of that information, the providers claim themselves to be foreign and they hide behind the American constitution.' – Dutch Forensic Institute

Technological developments and services on the internet go fast. What is currently used in criminal investigations, can, over time, for example after entering IPv6 addresses or other technological changes and innovations, be completely different. In addition, technological developments and new applications also allow for ambiguity or conflicting regulations.

> *'Prepaid data are kept for one year. This is a special piece of legislation*[104] *that was already there before the data retention legislation. [...]. Now there is a debate about how to consider a prepaid card for laptops. That's internet and thus, according to the data retention act, retention is limited to six months. Is prepaid internet then subject to six months retention or is prepaid internet subject to retention of a year? The explanatory memorandum of the two laws contradict each other herein. [As telecom authority] we have said that everything should be retained for one year if it is prepaid. When legislation is contradictory, as supervisor you must give clarity.'*
> – Telecom Agency

> *'The future requires that the regulations on data retention be flexible enough to adapt to new developments.'* – Public prosecutor

> *'IPv6 which will soon be introduced, has many different features. It works slightly differently to the IPv4. I think that experts will need to have a good look [at the retention laws] to know how things should be stored. What will be stored and what not.'* – Dutch Forensic Institute

The experts in this study pointed out another problem, namely the growing shortage of IP addresses. The IP addresses that are being used now, are usually IPv4 addresses. For years it has been known that the availability of IP addresses would run out, and that alternatives should be sought after. Such an alternative was found in the IPv6 addresses. Apart from a number of other benefits, the IPv6 address is longer and as such it is possible to create more addresses.
Slowly, more and more network equipment is being adapted to IPv6, but it is unlikely that a full transition will be possible in the foreseeable future. Services on the internet that only support IPv4 require the user also has an IPv4 address. It is thus undesirable for a user to only use an IPv6 address. Now that IPv4 addresses are scarce, providers are inclined to allot them parsimoniously. This is possible by bundling multiple customers through Network Address Translation (NAT) to share IP addresses or to assign IP addresses to users for a very short time.
When internet providers group a number of subscribers such that they use a single external IP address, it is no longer possible to determine from where

104 Article 13.4 section 3 of the Telecommunications Act. This describes that both prepaid traffic data, as well as location data must be retained for a period of one year. This holds for both telephony and internet data.

certain information has been sent. A Telecommunications Research Information Service request will thus yield several hits. When providers have trouble linking a user to IP addresses however, a Telecommunications Research Information Service request won't even yield one hit. Moreover, if an IP address brought in conjunction with a specific time is no longer unique to a specific subscriber, only extensive data retention and further inquiry into these data can yield a mere indication as to the subscriber responsible for a specific message. Internet Service Providers are not required to keep track of the computers with which subscribers are connected to the internet. The scope of internet users is too large for such precise recording of data. In addition, internet providers are not required to retain other unique characteristics of internet traffic, such as the port addresses that are specific to an IP address. It is thus expected that, increasingly, an IP address will lead only to a large group of subscribers, from which investigative services will still have to determine exact users.

5.3.7 *Telecommunications Research Information Service IP address requests*

The Telecommunications Research Information Service data is refreshed every 24 hours. The database therefore has no access to information older than that. IP addresses are, however, re-issued regularly. This holds true particularly for mobile internet, where IP address used can change several times a day even. But even fixed home computers can switch IP address. A Telecommunications Research Information Service request therefore can provide unreliable information in such cases. The exact time when an IP address was active is essential to find the correct information. The respondents in this study indicated that sometimes mistakes are made in this regard.

> *'Requests are most often sent to the Telecommunications Research Information Service, but for IP addresses, that doesn't make much sense because the addresses change so frequently, so you end up having to make a series of requests directly to the provider. But that's a little complicated. It's not something you can do automated and on a large scale.'* – Police

> *'I'm afraid that a lot of IP address requests are still sent to the Telecommunications Research Information Service. The registered name of an IP address today is not the same as the name on the IP address of three weeks ago. And for that you actually need to know a little more about this. That's one way in which errors can creep in.'* – Police

> *'[...] that error, the one where the current database is consulted, is one that's been made more often. So this can lead to different outcomes. It is not complicated, but you need to be meticulous. The quickest database is obviously*

> *that of the Telecommunications Research Information Service. But that can lead to the wrong information.'* – Provider

5.3.8 *Retention periods*

As has been shown earlier in this report, internet traffic data is not frequently used. On the one hand, this is due to the dated regulations yielding decreasingly useful data. On the other hand, it is due to a strong knowledge deficit. Despite this, the professionals and experts interviewed who act in criminal cases involving internet data, indicate unanimously that the retention period of six months is too short. Mostly this pertains to the data concerning the identification of a user on the basis of an IP address.

> *'[...]. You always work in the past. Such an investigation comes up suddenly and you usually don't work in the here and now. And then if you only have half a year, then that's very tight.'* – Fiscal Intelligence and Information Service

> *'[...] The more serious the crime, the longer the investigation period will be, and the sooner you'll find yourself in the situation that you want to request data from longer than six months ago.'* – Public prosecutor

> *'At the time that I arrest someone, the data that he has now are current and a half year seems like a long time. But people seem to keep their whole lives on the computer and that means that for a lot of information, you still don't get very far.'* – Police

Several professionals and experts provided examples of investigations where internet traffic data wasn't available and had an adverse effect on the investigation. A Fiscal Intelligence Investigation Service official explained that tax returns are always done at the end of the (tax) year but that fraud can only be determined and further investigated after the tax returns have been filed. In one case of childcare benefit (a personal supplement that can be requested only for own children) fraud, an application could be submitted through the internet. During an investigation, the Fiscal Intelligence Investigation Service found that, using false identities and a few IP addresses, childcare benefits were applied for an improbable number of children. From the time of this discovery, up to six months prior, it was possible to obtain the IP addresses. However, some of the applications had been made prior to that and it thus wasn't possible to obtain the historic internet traffic data. Other experts also cited examples of cases where the retention period of six months posed problems.

'[...] we conducted an investigation of the computer logs pertaining to conversations [on child pornography] and there were a lot of logs, stored over a period of several years. The suspect was arrested in December. I think by the time we had the data in our own system so that it could be analyzed, it was about March [...]. By the time it's March, and you only have a retention directive of six months, you can only retrieve data from October. Two months tops is all we had left to retrieve the IP addresses on the basis of the latest logs, that could potentially lead to another suspect.' – Police

'I think that the retention obligation is out of balance because a lot of data are stored that are never requested, whereas the one thing that is requested a lot is available only for too short a period of time.' – Dutch Forensic Institute

Experts and professionals working in teams investigating longer-term cases, such as liquidations or organized crime, indicated that they would like to see internet retention periods to be the same as those for telephony data. In addition, several respondents note that it is actually not possible to disconnect smartphone traffic data from the internet.

'It's actually not possible to separate, because the data applicable to the mobile phone can't be seen as separate from the internet. If a smartphone was connected to the internet nine months ago, then the fact that that internet session took place doesn't have to be retained by the provider. But what the provider does have to store is the date and time when the device was used to connect to a network. That is what the provider does have to retain. Then you automatically get the next question "Nine months ago that smartphone did something that lasted half an hour and it occurred through the use of that antenna etc." Luckily it doesn't work that way, because the provider only says it was an IP session, not how the session took place. The provider can't deliver any information anymore concerning how the IP session took place, from which location, with which service or using which IP address.' – Police

5.3.9 Requesting international traffic data

When investigators wish to access traffic and location data from non-Dutch persons or from non-Dutch communications service providers, an international data request can be made. As mentioned earlier, respondents reported not making international requests for legal assistance often, because of the length of time it takes for that data to become available. How long it takes varies according to the respondents. Sometimes, international colleagues are contacted right away, to 'freeze' the requested data in advance, ensuring that data doesn't get 'lost' due to the delay carried by a the international request

for legal assistance. The formalities of the international legal assistance request are then preceded by a telephone call to a colleague abroad. International cooperation is essential because the Telecommunications Act stops at the Dutch border.
An expert specialized in combating internet crime, indicated that his team makes little use of the data stored domestically. This has little to do with a lack of knowledge on internet data, but rather with the way the suspects work and their nationality, if they can even be identified. According to the same expert, in cases when the investigation team does want access to traffic data, it generally is requested from abroad. In that respect, investigators are thus dependent of the way in which the data retention directive is regulated and implemented in other countries, and generally investigators don't know in advance which data will be provided.

> *'Due to our size, for our own investigations with domestic suspects, we don't often request data meeting the specific retention requirements for one of the four or five suspects in the Netherlands. On the other hand, we regularly request data from abroad concerning the identity of a user or owner of an IP address at a certain time. This happens frequently vice versa as well.'* – Police

> *'[...]. It's a shame the terms aren't unified throughout Europe. That's a bit of a missed opportunity [...] now it appears as if the different countries have different views about it. That's a shame.'* – Public prosecutor

> *'We often put things on to other countries, so the cases will be handled and closed there. For example, in Germany, the problem is that they hardly have a retention requirement. What you have at the moment is what you have, but if it's not in the computer, then you can be sure you won't have it. At this moment, especially when you consider the new child pornography organization – there are eleven teams in the country with 150 people working on it – this is just a very important issue [identifying a user behind an IP address] to be able to achieve something.'* – Police

When traffic data are required from foreign Internet Service Providers such as Google, Facebook, Microsoft and Apple, an international request for legal assistance is needed in the Netherlands to do so. The services provided by these companies do not fall under the Dutch Telecommunications Act, chapter 13, because the parent company is located abroad. During our interviews, it regularly came up that respondents felt they were lacking information from frequently used internet services like Hotmail, Gmail, WhatsApp, Skype, Live Messenger, and so on. There are no figures concerning the frequency of Dutch international data requests for social media services. However, in light of the Public Nature of Government Act (*Wet Openbaarheid van Bestuur,*

WOB) request made by Bits of Freedom, the Limburg-Zuid police have in the past, made public a list of data requests from Hyves, Hotmail and Microsoft's Live Messenger.[105] In 2011 there were 33 requests made from Hotmail and Live Messenger, and six requests for information from Hyves. The latter company is in fact a Dutch provider of internet services.

The figures for the number of data requests made of Google have been published on the internet.[106] The services that this company provides include Gmail, Google Groups, Google Talk, Google Voice, etcetera. The company itself deals with data requests in the context of criminal investigations, checking to see if all requirements have been met. Data requests are to be in line with both the letter and the spirit of the law.[107] In the second half of 2012, 59 data requests from the Netherlands were processed by the company, of which 76% were complied with. It is unclear whether these responses were complete, who the applicant was and for what reasons or on what legal grounds the data requests had been made.

Microsoft has also publicized information regarding the number of data requests made by investigative services in 2012.[108] The online services provided by Microsoft include, amongst others, Hotmail, Outlook.com, Skydrive, Xbox Live, and Skype. In total, 859 data requests were received by Microsoft from Dutch investigative services, concerning 1,438 accounts and/ or users. None of the data requests yielded contents of the communications. However, 78.1% of the data requests were answered with traffic data being provided to the investigative services. In 21.9% of cases, no data were found or the request was denied due to not meeting the legal requirements. Skype received two data requests from Dutch investigative services, pertaining to two users. These requests were also denied. For Twitter, the number of data requests made by Dutch investigative services in the second half of 2012, was less than ten.[109]

It is noteworthy to mention that the frequency with which Dutch investigative services made data requests to the above mentioned foreign Internet Service Providers, is only a fraction of the frequency with which neighbouring countries such as Germany, France and The United Kingdom requested data.

5.3.10 The future of data retention for internet data

Increasingly, internet has been gaining a greater share of people's communication flows, and this trend is continuing. It has become clear, from the previous sections, that internet traffic data stored under the data retention directive has only limited value. One reason for this is the fact that technology,

105 www.bof.nl/live/wp-content/uploads/20120217-bevragingen-sociale-netwerken.pdf (consulted on March 19, 2013).
106 www.google.com/transparencyreport/userdatarequests/NL (consulted on March 19, 2013).
107 www.google.com/transparencyreport/userdatarequests/legalprocess (consulted on March 19, 2013).
108 http://blogs.technet.com/b/microsoft_on_the_issues/archive/2013/03/21/microsoft-releases-2012-law-enforcement-requests-report.aspx (consulted on March 25, 2013).
109 https://transparency.twitter.com/information-requests/ttr2 (consulted on March 25, 2013).

human behaviour and legislation don't seem to be suited to each other. The question then remains: 'Which information should be retained in order to best serve detection and investigation authorities?', and following from that obviously: 'Is it feasible?' and: 'Is it desirable?' Though retaining more data may seem to be a logical step, it is questionable whether or not this is a desirable solution. It is, in fact, a complex issue requiring careful consideration of the experts.

> *'An important discussion in the development of the retention directive was, of course, Appendix A for telephony. This refers to the external characteristics of the traffic, when does a number call another number, how long does it last and from which location was the call made. For internet, that is far more difficult to distinguish. It is difficult to distinguish between the external characteristics and the content of a message. Frequently you're in all the different levels of the Internet Protocol and you need to be in the deepest layer to actually have any use for it. In that sense, the content of a communications then becomes part of what you would need to retain, and that is a big step to take, of course. Retention of a year would be quite exceptional in terms of storage capacity, but secondly, that would be the ultimate Big Brother.'* – Provider

> *'[...] A Wi-Fi point at the supermarket has a certain assigned IP address, so then we [the provider] have to hold connection from point A to point B to point C. In order to demonstrate that, we must be specific in the time, irrefutably correct. You have to be able to prove in court that it is a correct record. [...]. In order to do that, we have to record the entire communication. Record much more than we do now.'* – Provider

5.4 The retrieval of transmission tower data

The data retained under the retention directive can be requested and investigated in different ways. Besides retrieving historical traffic telephony and internet data from individual numbers, IP addresses or devices, traffic data may also be requested from a particular location. For both telephony and internet traffic data, the location of the start of the connection, the so-called First Cell ID, is stored. These data refer to a transmission tower from where the connection originated. In other words, when a group of people start a call via the same transmission tower, all individuals will have, apart from their personal phone number, identical location data (First Cell ID) in their traffic data referring to that particular transmission tower.
A data request from a particular location is formulated in such a way that all communications from tower X at time Y are selected from that particular database. The retrieval of traffic data based on a location yields the data of all

mobile phones that have been called, or that have made a phone call themselves or a connection with the internet via that particular transmission tower for the requested time frame. The conditions that need to be met in order to warrant such a data request are that there must be a suspicion of a crime as described in Article 67, section 1 of the Code of Criminal Procedure, and that the data must be relevant to the investigation.
In addition to the legal requirements imposed on the requesting of traffic data from a transmission tower location, there are also agreements between providers and the National Interception Unit. For example, there is a limit to the number of transmission tower sites that can be requested per data request as well as to the time period over which data can be retrieved. This time period is limited to a maximum of three hours per tower location.

5.4.1 In practice

By requesting traffic data from a certain transmission tower location, it is possible to gain insight into the telecommunications traffic within a certain area, which can be useful for detection and investigation. The probability that the retrieval of location data will yield any useful information is the greatest when there is a specific direct reason for that request, like finding out if a communication took place using the suspects phone at the crime scene. The traffic data generated by communication are subject to the retention act and are thus available to be requested by investigative services for a certain period of time.
Mobile phones make regular contact with the network, even when no phone calls or data traffic occur. This type of data does not fall under the retention act and is thus only available for several hours before being overwritten. Still, these data can be claimed by investigative authorities under provisions Articles 126*ng/ug* jo., 126*nd/ud* and 126*ne/ue* of the Code of Criminal Procedure. In order to avoid data being overwritten, the data can be 'frozen', after which these frozen data can be claimed. However, according to one of the respondents, not all providers are able to 'freeze' data, ensuring its availability for disclosure claims.
In what cases investigation teams request transmission tower data, and how does this work? Below we have outlined several examples.
A murder has been committed at a particular location and the perpetrator is unknown. A possible detection operation could be to determine whether the perpetrator was in procession of his phone and if that phone made connection with a transmission tower in the vicinity of the crime scene. In that case a police telecommunications specialist does the relevant readings. This specialist accompanies the detectives from the investigation to the crime scene, where the probable route of the suspect is traced on the basis of tracks or witness statements. The coverage of the transmission towers are measured in the places along the route, as well as, of course, at the crime

scene itself. Most towers have three antennas each covering an area of 120 degrees. These areas covered by a transmission tower are referred to as 'cells'.[110] Though for multiple transmitters within a given area, overlap may occur, generally there will be a dominant transmission tower in a cell, which may not necessarily be the nearest location. Mobile devices seek connection with the dominant tower in a cell to enable communication. Good measurements conducted by a telecommunications specialist are thus necessary to determine for which tower location data should be claimed, as well as to determine the time frame for which to request relevant data. The retrieval of data from one location yields phone and IMEI numbers that have connected with that particular tower, which in turn can be claimed by the Telecommunications Research Information Centre. The investigative team can then sift through the data to see if there are familiar numbers on the list, such as those belonging to sex offenders, repeat offenders and so on. However, this method is not preferred due to its laborious nature, and is thus only done when there are no other useful clues.
Another method to determine the identity of a suspect is to compare transmission data derived from cells in different locations. If two or more crime scenes are suspected, a specific route to the crime scene, of a suspect who has been active in more places at different times, as in the case of serial offenders, transmission tower data can be requested and compared from several locations. In that case, measurements are taken at the locations in question and data is claimed for the best serving cell, the cell with the strongest range in that location. The data from the transmitters at different locations are compared using analysis software, and the phone numbers occurring more frequently in the data set are singled out for further investigation. If the same phone or IMEI number is found in both locations, it is deduced that the phone may belong to a possible suspect. For example, one interviewee explained:

> *'If someone is shot here, and a half hours drive from here a car that can be linked to the case is set on fire, then you can compare the data from the transmission towers to see if someone can be placed at both crime scenes. That could be an indication that someone has some kind of involvement with the case.'* – Police

It is also possible to investigate whether two separate crimes were committed by the same offender. For example, a robbery at location X with a certain modus operandi can later be compared with a robbery at location Y with a similar modus operandi, in the hope of identifying one and the same suspect. If there is a reason to believe that a call has been made, a comparison of the traffic data is made between the corresponding transmission towers on the

110 For a more extensive explanation see https://rejo.zenger.nl/focus/locatie-te-achterhalen-uit-call-detail-records (consulted on May 1, 2013).

various locations, in order to investigate whether there are matching numbers. Quite some time can pass between the retrieval of the traffic data on the basis of a location of a transmission tower and the comparison of the data set derived from multiple cells. For instance, in cases of regular arson, traffic data can be claimed and stored by the investigative team for comparison to other fires at a later date. A prerequisite is however, that the sites are far enough apart. If not, then a large number of mobile phones, for example, belonging to the people living in that area, come up in analysis. Transmission data can also be used in a later stage of the investigation, to be compared with a seized mobile phone.

> *'There was this guy who was probably lying in wait for his victim for four hours before assaulting and raping her. We did transmission readings. Later it appeared that while he was waiting in the bushes there for four hours, he had received a text message and had had an internet connection. [...]. You can request the transmission data, but you get a huge bulk of data. [...]. In this, we arrested a suspect later and then seized his phone. Then his phone was read out and the historical traffic data was requested, where we could see a match between the cell ID in the traffic of the suspect [location of the phone] and that measured by our cell IDs [location of the phone mast].'* – Police

During our interviews, it became apparent that with the retrieval of data from a transmission tower location, sometimes witnesses can be identified as well. It is possible that, for example, several witnesses try to call 112 (the emergency service number) simultaneously. Only two or three people then get in touch with the emergency services, while other witnesses abort their phone call, either because they see others are already in touch with the emergency number, or because they simply can't get through. For the police, such witnesses can be of great importance to the investigation. Though on the borderline of the law, one respondent provides an example of the following incident:

> *'Here in [place name] we have a nasty neighbourhood, where an Antillean was shot in broad daylight in the middle of the street. There was quite a commotion and lots of people called 112. At the emergency room they took three of the 112 callers. Then you go to transmission tower histories, and then it turns out that there are twelve 112 callers among them, and now you have their phone numbers so you can call them right away. [...]. They are important witnesses. It is a bit on the edge of the law. It is usually only an extra. [...].'* – Police

5.4.2 Privacy

The retrieval of data based on locations of transmission towers is controversial. Opponents argue that the retrieval of data from a location violates the privacy of large numbers of innocent civilians. Police officers themselves also recognize the potential magnitude of the infringement:

> *'I personally think that huge violation of privacy is made when traffic data is claimed form transmission towers, because 99.9% of the people concerned are not suspected. It's not so much the term suspects in the way intended by Article 126, but these are people who you really have absolutely nothing to do with. Sometimes you have a very good reason to do so, but sometimes reasons given make me wonder whether we should do that and if you can explain yourself, and if it's proportionate.'* – Police

Proponents of transmission tower data retrieval claim, however, that the privacy infringement is limited. The retrieval of data based on the location of a transmission tower is subject to requirements of proportionality and subsidiarity. There must be an offense for which pre-trial detention is possible. The professionals we interviewed realized it to be a serious investigative tool, and they wanted to prevent this from becoming a regularly used instrument. Nowadays, there appears to be greater awareness as to the use of such serious instruments. For this reason it was agreed that data may only be retrieved over a maximum period of three hours. Accordingly, an interviewee provides an example that shows that transmission tower requests are not always necessary:

> *'Sometimes we only note the fixed antennas [pertaining to the scene] that are relevant and nothing else. No transmission tower data is requested then. Suppose we run into a suspect and his phone a month later, we then request the historical traffic data from his phone and we will see if there is a match between the measured antennas and antennas that occur in that historical traffic data [location data]. That way you can place his phone in the area of the crime scene at a particular time'* – Police

5.5 Alternatives to the retention directive?

We asked our respondents whether they could come up with alternatives to working with traffic data and the retention directive. Both professionals and the experts emphasize that the criminal investigation process is very dependent on telecommunications data. In their view, there is no viable alternative. If traffic and location data would not be available for investigation, the alternative would be to use wire tapping. This however, firstly poses a greater infringement of privacy, and secondly provides data about the future as opposed to historical data, which provides data about the past. According to the respondents, another alternative could be surveillance, but that requires surveillance teams that know where to stake out. Without that, they claim work becomes very inefficient. One expert pointed out that if he did not have traffic data available to him, he would consult other forms of data storage systems or closed circuit television (CCTV). But the conclusion remains that historical traffic data are very valuable and also relatively irreplaceable.

An alternative to data retention is freezing targeted data. In order to freeze data, investigative authorities are required to describe carefully which data set is to be retained, before the rest of the data are destroyed. The frozen data is also not supplied directly to the investigators, but only later when a disclosure claim is filed by the investigative authorities. General opponents of the retention directive regard freezing targeted data as less of a privacy infringement, because it entails retaining a well defined targeted dataset for a period of time, as opposed to retaining the traffic data of all of a provider's customers.

The researchers proposed this alternative to the professionals and experts in this study. None of the respondents thought that freezing of data was a comparable or equivalent alternative to retrieving traffic data stored under the retention directive. Freezing data is already based on a retention period, and the current retention directive has just ensured harmonization of retention periods among the different providers.

Another example of cases where freezing data isn't effective is when it only later comes to light that a crime has been committed. In the current situation, that is already the case for the fleeting tower data, stored only very briefly. As noted earlier, such data are only available for a few hours and can therefore only be claimed if the police is informed about an offense quickly, and if they also know then where the crime was committed. In other cases, the freezing of data is not a viable solution.

> *'You can ask a provider to freeze the short term data. That means that a snapshot is taken of the data from such a tower. That only works if you immediately become aware of a particular offense, for example. If you find out now that a murder was committed fourteen days ago, then you can freeze data until you're blue in the face, but everything is gone anyway.'*
> – Police

5.6 In sum

Crime investigators still hold on firmly to traditional methods of detection, such as historical telephony traffic data. These data are used frequently in a wide range of offenses. Our interviews showed that historical traffic and location data are used frequently for investigative purposes, and in very diverse ways. Professionals in the criminal investigation practice are familiar with the process and find it a valuable investigative tool. Locating persons is a frequently mentioned reason to claim data under the Telecommunications Data Retention Directive. By requesting traffic data, it can easily be determined whether, and from what location, a phone was used, providing potentially both incriminating and exonerating evidence. Traffic data are also frequently used to provide insight into contacts. Analyzing traffic and location data is specialist work, according to the experts interviewed. However, not every team has access to an analyst who can perform the task.

When requesting transmission tower data, that location is searched for in the traffic and location database. In this way, telephone numbers that were used to communicate at a given time and a particular location are selected. This type of data request is mainly used for the investigation of serial offenses, to enable a comparison between phone numbers and locations, in the hope that a suspect has used his phone near the places where he/she committed crimes.

According to the experts, the retained telephony data as described in the appendix to Article 13.2*a* of the Telecommunications Act, are relevant. Respondents did indicate that they missed certain information that is not automatically supplied with requested traffic and location data. An example of this is the final location of a call. Under the current conditions, only the starting location of a connection is traceable from the requested data. In general, the retention period is considered to be sufficient by investigation professionals and experts. In some cases, a year is too short to conduct a good investigation, but by and large that is the exception rather than the rule.

Professionals and experts who frequently investigate crimes involving the internet, believe that the retained internet data is only of limited value in investigating these cases. The retained data, as prescribed by the Telecommunications Act, no longer matches current technology and current use of internet. As a result, data of citizens are retained without ever, or frequently, being requested by the investigative services.

This study also highlighted that the technology behind internet data is complex and that the use and analysis of this data requires specific knowledge. That knowledge is not sufficiently available, according to those interviewed, despite attempts by the public prosecutor's office and the police service to overcome this deficiency.

Generally speaking, the six month retention period for internet traffic data is considered unanimously to be too short by our respondents. This mainly concerns the identifying data of a user of an IP address.

A careful review of the rules governing the internet data to be retained, and the retention period thereof, is therefore desirable.

6 The use of historical traffic data in figures

6.1 Data requests from the National Interception Unit

One of the rules stipulated in the Telecommunications Act, directs the publication of the annual number of telecommunications traffic data requests made by investigative agencies (Article 13.4 section 4, Telecommunications Act). The European directive also mentions the annual publication of the number of data requests per year (Article 10 of the Directive 2006/24/EC). The European evaluation however revealed that few member states fulfil this obligation. In those cases where figures have been published, the format and type of data is varied so that the figures are difficult to compare. Recently, a paper was published by the Platform for Electronic Data Retention for the Investigation, Detection and Prosecution of Serious Crime for the member states, which can serve as a guide for publishing their annual figures.[111]
In 2010, the Minister of Security and Justice disclosed the number of claims under the provision of telecommunications data for the first time. In the second half of 2010, the number was 24,012. In the year 2011, there were 49,695 claims filed. The total number of claims filed in the year 2012 amounted to 56,825. This is an increase of 14.3% compared to 2011.
It must be emphasized that, in the Netherlands, telecom data requests are registered per telephone number, IMEI number, IP address, or 'pole location', for which data is requested.[112] Because people often use multiple phones, these figures do not give insight into the number of people for whom telecom data have been claimed annually, nor into the number of investigations or the nature of the investigations for which the data is sought. These figures also don't provide insight into the extent to which a data claim actually led to the provision of that data. Only the number of claims are recorded, not the number of times that data is actually delivered.
For the period before 2010, no reliable figures are available because the registration of disclosure claims did not go through a central organisation at the time, and the police corps sought contact with providers individually to request data provision. In 2010, this situation changed: since then all actions concerning the provision of data, including historical traffic data, are processed by the National Interception Unit of the National Police. All incoming claims are centrally registered there.
Upon receipt of a claim, the National Interception Unit checks if the provider's name is correct, if the provider can deliver the requested data, the time period over which data is being requested, if the data requested meets the terms and conditions made by the various parties, and whether the request can be deemed reasonable. For the latter the check cannot be substantive, as the National Interception Unit does not have any information concerning the case itself, but it can check to see if the data request is

111 http://ec.europa.eu/dgs/home-affairs/what-we-do/policies/police-cooperation/data-retention/experts-group/docs/guidance_on_statistics_position_paper_16_datret_final_en.pdf (consulted on March 20, 2013).
112 For disclosure claims pertaining to smartphones, providers supply both information concerning the telephone number as well as concerning the IP address.

feasible. The claims are forwarded by the National Interception Unit to the appropriate provider, who in turn sends the requested data – if available – back to the National Interception Unit. The National Interception Unit sends it on to the police corps where the data request originated.
The total number of claims for disclosure in the year 2012 amounted to 56,825. For the first time, the figures of 2012 have been broken down by the categories below, which we will go through and describe, one by one. However, the number of claims as publicized by the Minister contain some data provision claims not covered by the retention act. For as much as possible, these have been kept out of Table 1 (in Section 6.1.1). Claims that the National Interception Unit placed in a 'residual category' have therefore also been excluded from Table 1. Data claims in this category include, for example, requests for mailbox dumps, whereby not only the traffic data is sought, but also the content of mail traffic. Therefore, the total number of claims in the table on page 79 is only 54,813 requests. Because this breakdown was not possible earlier, it is impossible to determine more specifically which category was responsible for the increase over previous years, or what the increase was caused by.
In 2012, 41,658 claims were filed pertaining to 'historical telecommunications traffic data'. This includes information on telephone use and possible IP traffic, such as with which number was dialled, when was called, what the duration of the call was, from which location was called, and if a connection was made with internet. When the traffic data is claimed for a smartphone, both telephone traffic data is supplied by the provider, as well as IP traffic data. Where disclosure claims were made only for IP traffic data, and no telephony traffic data were requested, they will be registered under 'Historical IP traffic data'.
Of the total number of disclosure claims for historical telecommunications traffic data, 42.6% was not older than three months. The average period for which access is sought is 27 days. The number of claims for historical traffic data from four to seven months old, is 9,487, which is 23% of the total. The average period over which access is then requested is 97 days. It is worth mentioning that the older the data that is requested, the longer the average period for which data is requested.
As described in Chapter 4, the researchers saw an overview of the historical traffic data of a smartphone, containing IP traffic data that was older than the six month retention period for IP data. It was not possible to determine whether the IP data in this category of this table, were or were not destroyed on time. In total, there were 3,376 data requests (8.2%) dating further back in time than the maximum retention period.

As can be seen in the 'Claims from transmission towers' category, 6,361 traffic and location data claims were made. These refer to claims where providers had been requested to supply data concerning which phones connected

with a transmission tower to establish communication from a particular location (the cell ID).
These claims, which are focused on the location of traffic data (the cell ID), provide an overview of all mobile calls made, that were initiated through the connection with a particular transmission tower.
A claim for traffic data based on Article 126*n* Code of Criminal Procedure always involves one number or identifying characteristic. Only in the case of transmission tower claims this is different, because that involves the retrieval of multiple cells which can show an overlap in their coverage of a particular area. However, transmission tower data claims are only counted once in the statistics. An exact figure on the number of transmission tower disclosure claims would provide a distorted image, because they often involve large numbers of mobile phones, which ultimately, after a thorough, often automated, analysis, yield a small portion of interest for further investigation.
More than three quarters of the number of transmission tower claims (79%) don't go further back in time than three months. In mutual agreements between the National Interception Unit and the police corps, it was decided that claims for traffic data from transmission towers would not exceed a period of three hours. This explains the requested average period of 0.2 days. However, it appears that when data requests go further back in time than three months, they sometimes do exceed the three hour limit. The National Interception Unit explains that in exceptional cases, the three hour rule can be deviated from, which explains the longer than three hour intervals.

'Historical traffic data email' and 'Historic traffic data IP' were requested to a very limited extent. Historical email traffic data was requested 213 times, historical IP traffic data 39 times. A possible explanation for this lies in the fact that only Dutch providers are required to retain email traffic data for investigative services. The added value for the retrieval of IP traffic data consists of obtaining the log-on and log-off data (see Chapter 5, section 5.3.1) that are then delivered. However, these data prove to be of very little value to investigative services. When it comes to traffic data that originates from a smartphone, requesting 'Historical traffic data' is the more obvious choice, thereby obtaining data to gain insight into both telephony and internet. The IP traffic data in this category doesn't provide an answer to the question which IP address someone was using. That type of claim is categorized under the heading 'Historic personal data' (the last category). Interestingly, both for the retrieval of historical email traffic data as well as for the retrieval of IP traffic data, a high percentage of disclosure claims extended beyond the six month expiration limit for this type of data.
The data claims in the last category of Table 1, 'Historical personal data', can pertain both to telephony as well as to IP data requests. This category can cover questions such as 'By whom was this phone in use two months ago?' or 'Who was using this IP address five days ago at 15:02 pm?'. The number of

requests in this category totals 6,542, of which 64% is not older than three months. The data requests in this category are very targeted questions. The exact time of a phone number or IP address use is of crucial importance for these types of data requests. This is reflected in the fact that the average period over which data is requested only amounts two days.

An important note to be made concerning the figures in Table 1, is that the claims they represent do not *only* cover disclosure claims subject to the retention act. The National Interception Unit is not able to generate statistics pertaining only to disclosure claims that fall under the retention act. For example, concerning companies and providers that offer services on the internet but have no storage requirement, it is possible for investigators to submit claims to obtain information relevant to detection. Administratively, these claims fall in the same category as disclosure claims for data that does fall under the retention directive. For this reason, disclosure claims of which it was clear that they are not covered by the retention act, have been excluded from Table 1, explaining the difference between the number of claims for data provision published by the Minister and the total number of claims. It is not clear how many claims not falling under the retention act are in fact included in Table 1.
Incidentally, this could provide an explanation for the fact that quite a number of claims fall in a period after the expiry of the retention periods. Another explanation could be that possibly due to strict administration rules, claims which aren't registered immediately, are ultimately registered in the next column of the table. In addition, no distinction is made between subscribers and prepaid users, despite the fact that prepaid IP data have a retention period of one year according Article 13.4 section 3 of the Telecommunications Act. Finally, it is also possible that the number of claims, according to these figures, that fall after the expiration of the retention period, actually indeed do pertain to claims whose term did exceed the maximum period of six months/one year. In that case, the claim was registered by the data request period, which does not necessarily mean that the provider complied. If the provider indeed destroys data according to the rules after the retention periods have expired, then this information would no longer exist. But people are free to ask for data over a period extending the retention period required by law, and investigators may do so. On the basis of this study, it is impossible to say whether providers still store data beyond directed retention periods, and whether providers deliver such data to the National Interception Unit in those cases. Unfortunately, the available figures do not provide insight into the extent to which providers answer the data requests made. The figures on providers' complying are not registered. A question such as 'Who was the user of this phone three months ago?' can be answered by the provider with 'It was Mrs. Smith' or 'Unknown'. Both answers are considered

as a data delivery. This is because the National Interception Unit forwards provider responses without looking into them.
The figures also do not provide insight into the annual number of persons whose traffic data have been requested. Criminals often have several (mostly prepaid) phones and SIM cards in use. Prepaid IP data are required to be kept for a period of one year (Article 13.4 section 3, Telecommunications Act), as opposed to the retention period of six months for subscribers. This distinction is however, not taken into account in the registration of data requests. When someone is the subject of a criminal investigation, data can be requested for multiple phones, IP addresses or email accounts. Therefore it is not possible to say to how many individuals the figures presented in Table 1 relate.
Moreover, the number of data requests is not counted double as is the case with the publication of annual figures for telephone and internet taps. When both telephone and IP tap are done on a smartphone, this is registered as two taps in the administration of the statistics. This is not the case with traffic data. When historical traffic data are requested for a smartphone, which yield both telecom and IP traffic data, that request only counts once in the statistics and falls under the heading 'Historical traffic data telecommunications'.

6.1.1 Conclusion

In general it can be concluded that the majority of the claims relate to traffic data that are up to six months old; a quarter of the claims pertain to data requests going back more than six months in time. In order to get a better idea of the exact number of annual data claims done, it is important that the management of the National Interception Unit adapt their registration to generate more reliable figures. Additionally, Dutch investigative authorities could provide more insight into the degree in which the Dutch authorities infringe upon the privacy of suspects, and others involved, using this detection method. They could accomplish this by, for example, registering how often a traffic data claim indeed leads to the delivery of the requested data. The publication of the Platform for Electronic Data Retention for the Investigation, Detection and Prosecution of Serious Crime could serve as a guide to publishing annual figures.[113] There it is recommended, amongst others, to count the number of negative answers and answers that don't provide any of the requested data. Furthermore, more insight could be provided by registering disclosure claims in a way that reveals how many people are subject to telecommunications traffic data requests each year, for how many cases this happens and for what type of cases these requests are made.

113 http://ec.europa.eu/dgs/home-affairs/what-we-do/policies/police-cooperation/data-retention/experts-group/docs/guidance_on_statistics_position_paper_16_datret_final_en.pdf (consulted on March 20, 2013).

Table 1 **Total number of disclosure claims for telecommunications data 2012**

Figures 2012		Retention periods broken down									
		0 to 3 months		4 to 6 months		7 to 12 months		13 to 24 months		>24 months	
	Total	Number of requests	Average period data requested in days	Number of requests	Average period data requested in days	Number of requests	Average period data requested in days	Number of requests	Average period data requested in days	Number of requests	Average period data requested in days
Historical traffic data telecom-munication	41.658	17.754	27	9.487	97	11.041	216	3.298	352	78	1.309
Claims from transmission towers	6.361	5.004	0,2	952	0,5	200	1,1	18	284	0	0
Historical traffic data email	213	68	22	85	96	42	131	18	350	0	0
Historical traffic data IP	39	14	11	10	27	14	115	1	169	0	0
Historical per-sonal data	6.542	4.167	2	1.349	37	696	84	230	191	100	659

6.2 The use of traffic data in jurisprudence

In this section we will discuss if and how Dutch courts use traffic data as evidence. This study included a selection of court rulings, because previous research done by Mevis et al. (2005) focused primarily on police data and records. The role of traffic data in court rulings is only briefly discussed in this study.[114] To address this shortcoming, criminal court rulings published on the website *rechtspraak.nl* were searched using search terms as '(historical) traffic data' and related terms that indicate the use of traffic data, such as towers, tower data and telecom data. The search term IP address was also used.
The website rechtspraak.nl contains a large collection of published court rulings from the District Courts, courts of appeal, and the Supreme Court. There are several criteria by which is determined whether a court ruling is published on *rechtspraak.nl* if not even mandatory. The latter applies in any case – insofar relevant to our study – where the rulings concern criminal cases of indictments (partly) for life threatening crimes or where in the court ruling yields an unconditional prison sentence of four years or more and/or a Custodial Psychiatric Care (*TBS*) measure has been imposed. In addition, publication follows for cases that have received media attention and when publication has value for legal purposes. Thus, in the set of court rulings that's published on the website annually, publication of court rulings for serious crimes is likely overrepresented.
There are also other reasons why the selection of court rulings used in this study is not representative of the cases that are brought before the courts in which traffic data are used. The fact that traffic data is requested for detective investigation, or that they have played a role during the investigation process, is often not included in a court ruling. For example, if a suspect confesses to the police or courts, or if other convincing and conclusive evidence is present, often the traffic data is no longer needed as evidence and is, despite it constituting incriminating evidence, not necessarily mentioned in the court ruling, even though its role in detection may have been crucial. The same applies to exonerating evidence and acquittals. The issues discussed below are all related to suspects who (partly) deny the charges against them.
Despite these limitations, our selection of court rulings is interesting because it provides insight into the evidentiary value attributed to this type of information by the court, and it shows what role traffic data can play in court judgements.

In this chapter we therefore provide insight into the use and value of traffic data in judicial rulings, and implicitly in the role of traffic data in the criminal investigation process, because what we find in the law is a reflection of the

114 See earlier the brief overview of court rulings for the purpose of a dialogue with Parliament.

investigation.[115] However, it is not possible to say anything about the time and the period for which traffic data are requested on the basis of the court rulings. We can thus not say whether it pertained to data dating back a few months, a year or even longer, because this information generally isn't included in the court ruling, and consequently cannot be reconstructed reliably.

In addition, we would like to note here that the selection of cases used, predominantly represents those where traffic data was used successfully. That is obvious, because it concerns a selection of cases which have been prosecuted, which implies that the public prosecutor considered there to be sufficient evidence to bring the case to trial. Investigations of which traffic data were requested (in combination with other detection methods) but that didn't lead to the identification of a suspect, or to trial, are not included in this selection. In some of the cases below, the courts ruled that the traffic data was insufficient to contribute to a conviction.

In this chapter we will present the results for the search terms 'historical data' and 'IP address' successively. At first glance, it can be said that the number of hits with the term 'IP address' was significantly lower than with the term 'historical (traffic) data'. For this reason, the period over which we searched the term 'IP address' in court rulings was longer than for the term 'historical (traffic) data. With regard to the historical term (traffic) data views, court rulings published between July 2012 and February 2013 were searched for the term 'IP address' the period ran from January 2009 to February 2013. To give an impression of the total number of published court rulings relevant to our research, in which we searched the term 'historical (traffic) data', we first present the following table (Table 2).

Table 2 Number of criminal cases published from July 2012 to February 2013

Court	Number of criminal cases
District Court	2.437
Court of Appeal	839
Total	3.276

From this, a further selection was made. Due to the less serious nature of crimes tried before the magistrate, and thus the diminished likelihood that traffic data would be used, these cases were not included. Moreover, generally magistrate's rulings are not written. Of those cases tried in the District Courts, only those tried by a panel of judges were selected here, and of the cases brought before the Court of Appeal, only the higher Appeal.

115 See Chapter 5 in particular, for the role of traffic data in criminal investigations.

Court	Number of criminal cases
Panel of judges	2.344
Court of Appeal (Appeal)	823

The cases that were not relevant for our study, pertaining to Custodial Psychiatric Care cases (28 District Court cases, and 149 Court of Appeal cases), including term extensions therefore, can also be deducted from the figures. This leaves the following selection of cases.

Court	Number of criminal cases
Panel of judges	2.195
Court of Appeal	795

In other words, in total, 2,195 District Court cases and 795 Court of Appeal cases were searched for (historical) traffic data and related terms that indicate the use of traffic data, such as transmission towers, tower data and telecom data as well as for the terms IP address and IP data.

6.2.1 Telephony traffic data

We searched for historical (traffic) data in the register of court rulings published on rechtspraak.nl. A total of 74 rulings in which the term historical data occurred were found. In these rulings we also found terms such as tower data and tower location. Upon further analysis, we found that the value, or function, of the data as evidence could be distinguished further to show different types of usage.

The categories that are most common, can be summarized under the heading 'Contacts between defendants' and 'Determining location'. By determining location we mean that, based on transmission tower data, a particular place was located where a mobile phone was used, it often being the crime scene, though not always. This is 'static' positioning. Besides determining location, we also distinguish a smaller category, namely 'travel movement', in which the route travelled by a suspect can be reconstructed based on several locations.

Establishing contacts between co-defendants and determining location often coincide, for example, in burglaries committed by more than one person, where telephone or text message contact is made between suspects, shortly before or after the fact.

In addition, we can see that, on the basis of historical data, it can be established that defendants have had contact with a victim or with third parties (i.e. witnesses). Moreover, we see that defendants' statements are explicitly contrasted with information concluded from historical traffic data. In the following table, the various functions of historical traffic data are shown. It should be noted that, in a particular ruling, different functional values of traffic data as evidence can be found. Firstly, a list of the different crimes with which the court rulings were related, (Table 3), subsequently an overview of the various functional values as evidence for traffic data used (Table 4).

Table 3 Overview of offences where traffic data was used (in the investigation)

Offences	Number of times appearing in court rulings
Burglary	24
Robbery home/store	14
Murder/manslaughter	13
Drug dealing	8
Larceny with violence/mugging/bribery	5
Kidnapping/hijacking	4
Threat	2
Fraud	1
Ram raid	1
Arson	2

If we look at the function of the traffic data in these court rulings, then we can see that these data are mainly used as evidence that people have been positioned in certain places and that they have been in contact with certain others. Below is an overview of the functional value as evidence of these data. In some court rulings, a combination of functional values as evidence is involved (see Table 4).

Table 4 Overview of the functional use of traffic data

Functional use of traffic data	Number of times appearing in court rulings
Determining location	39
Contact suspect	24
Travel movement	19
Contrast in statement	12
Contact victim	6
Contact witness	3
Evidence other	10

Below we will discuss a number of court rulings roughly grouped around the different functions and offenses mentioned above.

6.2.2 *Localization of suspects or networks and establishing their contacts*

Traffic data appear to be used relatively frequently to prove the location of suspects in court rulings related to robberies. Several examples of this are listed below.

> *A suspect is sentenced to six years for two house robberies with accomplices. During the robbery, the suspect was provided information by phone, about where the money could be found in the house. The Court held that, according to the historical traffic data from the mobile phones of two defendants,*

and contrary to the suspect's statement, the defendant had had contact with accomplices at the time of the robberies. (ECLI: NL: GHSGR: 2012: BY1648)

A suspect is sentenced to four years and six months for a robbery committed with an accomplice at a toy store, where the owner was threatened with a firearm. From historical traffic data it was shown that the defendant and his co-defendant, whose fingerprints and DNA were found on a plastic bag left by one of the robbers in the store, had had phone contact twice, shortly before the robbery, whereby both phones used a transmission tower in the immediate vicinity of the accomplice's home. From this the court concluded that the defendant picked up his accomplice at his home shortly before the robbery. (ECLI: NL: RBSGR: 2012: BX6147)

The next example pertains to a home robbery with accomplice whereby the victim was threatened with a weapon, beaten and tied up. The suspect's phone makes contact with a transmission tower in the vicinity of the house. The court further noted that it was 'notable (...) that from the time of the robbery the phone was not used to make calls or send text messages, only made sporadic contact with the transmission tower.' (ECLI: NL: RBSGR: 2012: BX5776)

In several cases traffic data were used to trace the travel movements of suspects prior to a robbery. For example, in a robbery, the travel movements reconstructed on the basis of tower locations, were mapped from a suspect's residence to the home of the victim. In a *rip deal* with fatal outcome, historical and tower data were used to determine the route travelled by the suspect and his presence at the scene of the crime. (ECLI: NL: GHARN: 2012: BX6121)

In a violent raid on a house, where the victim sustained serious bodily injury, the suspect's route (relevant to the evidence) was meticulously traced using the time and duration of a phone conversation and the street name of the location of the transmission tower. In addition, the court found that the relevant telephone numbers were used by the perpetrators for the robbery, and that the phones were specially purchased for use in the robbery. (ECLI: NL: RBARN: 2012: BY2895)

Regarding suspects of a raid on a courier, the court held that, derived from the telephone traffic data, one of the suspects was in fact not where he claimed to have been. (ECLI: NL: RBSGR: 2012: BX5105)

Finally, we mention a robbery at a McDonald's restaurant.

Research of historical data showed that a certain number was in use by a suspect and that two phone contacts from this number were made shortly following each other from two different locations, between which the McDonald's was located. With regard to the suspect's defence, that he lived in the vicinity of the McDonald's, the court noted that 'that tower data in principle has a supporting character, but that the data considered in relation to each other and in conjunction with other facts and circumstances from the case, can give reason to evidence of the crime. (...) With regard to the robbery at the McDonald's (...) it was revealed that in the six months prior to the robbery, the suspect's phone didn't once make contact with the transmission tower.'

With regard to a second robbery, also of a cafeteria, it was also determined that, in light of the contact made with the transmission tower, the robbery roughly was located in the area between the two towers. This was viewed in conjunction with the times at which the towers were connected with and the time of the robbery. (ECLI: NL: RBALK: 2012: BX4768)

The court rulings studied also provided examples of the use of location data in establishing contacts between suspects in other crimes than robberies.

Here we will provide some examples of these types of cases:

A defendant was convicted of complicity in a liquidation that took place from a van in the parking lot along the highway. Both the suspects' travel movements, as well as their purchase of ski masks and gloves, were established by the traffic data. (ECLI: NL: RBUTR: 2012: BX2092).

In an investigation into an attempted liquidation, the police discovered a partially burned out phone on a gas stove. Research by the Dutch Forensic Institute into this phone and the historical data of the (apparently) outdated number thereof, revealed that the phone was part of a so-called closed telephone circuit. These phones are used only to call members of the closed circuit, generally to avoid being traced by (prepaid) numbers belonging to third parties. The historical data provide insight into the suspects' driving route and their presence in the immediate vicinity of the crime scene. (ECLI: NL: RBAMS: 2012: BX1952)

In a large investigation into the production of synthetic drugs, contact between suspects was established on the basis of a note with a phone number on it, found in a search. According to the historical traffic data, this phone number was known to have called a particular buzzer. (ECLI: NL: GHSHE: 2012: BW7042)

An attempted homicide in which the victim was stabbed in the neck with a screwdriver. The court assumes that the suspect was present at the crime scene, a garage, and substantiates this with the information that a telephone tower in the vicinity of the garage was connected with. (ECLI: NL: GHAMS: 2012: BY2562)

A suspect was convicted of kidnapping and raping a six year old girl. The victim was lured into a car where the abuse took place. The historical phone records showed that the suspect could be positioned in the area, on the day and time, where a man had spoken to the victim. (ECLI: NL: RBSGR: 2012: BY0109)

Another case involved a large group of offenders committing large number of burglaries and safe cracking in stores and supermarkets across the country. From the historical data, it was established that the suspects were present at the crime scenes at the time of the burglary, or before for exploratory purposes. (ECLI: NL: RBUTR: 2012: BX9634)

In an investigation of truck cargo thefts, historical traffic data and tower data indicated that two phone numbers in use by two suspects, connected with the transmission tower in the vicinity of the crime scene, at approximately the same time a transmitter placed in the vehicle of the suspect did. (ECLI: NL: RBUTR: 2012: BX9634)

The suspect's phone made contact with a tower in the area of a number of burglaries. In conjunction with, among others, the statement of a co-defendant, this led to a conviction for several burglaries. The co-defendant had stated seeing the main suspect the morning after, in the company of others, coming home with things in his possession; the same things that were later found in a car driven by the main suspect. (ECLI: NL: GHAMS: 2012: BY1810)

In addition to these violent crimes, we would like to mention the investigation into a suspect's scamming of several banks across the country. Transmission tower data were used to substantiate that the suspect was in the vicinity of, or en route to, the banks where the scam took place. (ECLI: NL: RBBRE: 2012: BX4244)

Finally, we would like to mention the conviction of a defendant for the smuggling of cocaine.

On the basis of historical traffic data, the Court established that the suspect used a particular phone which subsequently made contact with a transmission tower at a terminal at Schiphol. (ECLI: NL: RBHAA: 2012: BW2968)

The preceding examples show that historical and transmission tower data may be of interest for determining the whereabouts of suspects at a particular time. Where telephone contact occurred between multiple defendants, as was the case with several robberies and burglaries, such data can provide relevant information for detection. A call that is made at or around the time of a crime, even if that's a call concerning the delivery of drugs, increases the investigation possibilities for the law enforcement services.

6.2.3 *Supportive or refuting statements*

In the studied court rulings, we found more ways in which historical and transmission tower data are used as (supporting) evidence. For example, the court explicitly addresses the inconsistencies between suspects' statements, or lack thereof, with the traffic data included in the dossier. Though it is conceivable that a defendant is confronted with (certain inferences from) the traffic data, our concern here is to highlight the cases where the court ruling pays explicit attention to the valuation of such data in the context of statements made by the defendant. For example, a defendant's explanation as to the (incriminating) evidence from the traffic data, is thus tested and in some cases found implausible. Additionally, in some instances where defendants refuse to explain, when an explanation seems fitting, the evidential value of the traffic data appears to be increased.

Similarly, the court addresses several seemingly contradictory points in the case of a person who only makes a statement several weeks after the suspected embezzlement of a large sum of money from a money transport vehicle.

> *This suspect claimed to have given his phone to an acquaintance, while historical data from the phone showed that he had sent his sister a text message. When questioned about this by the police, the suspect refused to give any clarification. At the court hearing however, the suspect admitted to have sent the message but, in defence of his refusal to say anything when questioned earlier, he claimed to have wanted to protect his sister. The court didn't appear to take this claim very seriously, and states that the defendant was unable to make clear how his earlier inaccurate statements could contribute to protecting his sister. (ECLI: NL: RBROT: 2012: BX1291)*

> *In another case, the court considered a robbery suspect's explanation pertaining to the historical traffic data and location of his mobile phone implausible. The defendant claimed that he might have left his phone in one of his football friend's car, and that those friends might in fact have committed the robbery themselves, using his belongings. The Court noted the improbability of this, particularly because the defendant had never*

mentioned the football friends in earlier interrogations. (ECLI: NL: RBALK: 2012: BX4768)

In a case against a suspect in an armed robbery in association, the court asserted that the defendant gave different statements regarding his presence at a particular location. It was suggested that the suspect changed his statement when the investigation yielded new information. This was especially true for statements pertaining to the historical traffic data from both his phone and the accomplice's statements. The court therefore didn't attach any value to the defendant's statements denying his involvement in the robbery. (ECLI: NL: RBAMS: 2012: BX5674)

In the rip deal with fatalities mentioned earlier, the Court of Appeal held the defendant and co-defendant's statements did not match the historical print and transmission tower data. The haste with which the defendants supposedly had to leave in connection with a birthday party elsewhere, was not in keeping with these data which showed that the defendant had stayed for an hour in Arnhem. Even if the defendant would have stayed in Arnhem to eat some French fries, an hour long stay in Arnhem seemed unlikely without other activities. The duration of defendants stay in Arnhem was derived from the phone data, and the offense was committed in the Arnhem area. (ECLI: NL: GHARN: 2012: BX6113)

In a case of a violent robbery with serious bodily injury, the Court of Appeal pointed out that from the defendant's explanation for phone contact with his accomplices, the Court could conclude that defendant was in the immediate vicinity of the victim's home at least twice on the given evening. With regard to the defendant's statement concerning a particular moment at which he had supposedly been called, the Court noted that that statement was irrefutably false in light of the telecommunications. On the basis of both historical print data from the telephone contacts and the report of a burglary, it was established that the defendant and his accomplices were on the road together, moving about several times, until late in the evening. (ECLI: NL: GHARN: 2012: BW8652)

A mirrored case occurred in a study in which a defendant just wanted to see his statement supported by the telecom data of a co-defendant.

In a murder case, the co-defendant submitted a statement incriminating the defendant. In his defence, it was suggested that the police and Prosecution Service had conducted a one-sided investigation, to which the public prosecutor alleged that the authorities attempted to establish that the co-defendant and the victim had known each other, amongst others, by doing a telecom investigation. The mere fact that this investigation did not lead to evidence supporting the defendant's statement was not enough to support the allegation that the investigation had been one-sided. The court held that there were indications to check the suspect's statements, for example with the use of telecommunications data, this indeed was done. (ECLI: NL: RBAMS: 2012: BX3164)

6.2.4 Other functions of the use of traffic data

In the above summarized court rulings it was shown that the courts can conclude contact between defendants on the basis of traffic data, in conjunction with other incriminating evidence, and how this in turn can contribute to the body of evidence. In the court rulings we studied, we also found that telephone conversations between suspects and victims, or between suspects and other parties, were considered relevant for evidence. To some degree this was also the case in the earlier mentioned court ruling where the defendant's sending of a text message to his sister undermined his statement. In a case of manslaughter by the victim's partner, historical traffic data made it possible to make a timeline for the times when phone contact was made between the victim and a third party, and as of which time there were no more signs of life. (ECLI: NL: RBSHE: 2012: BY0575). In the investigation of a manslaughter case, where the body was not found, a meticulous reconstruction of traffic data and phone contacts, between the defendant and the victim, led to a conviction. (ECLI: NL: RBNNE: 2013: BY9376).

In a case involving a burglary and fatal shooting at a cannabis farm, the defendant was convicted. On the basis of historical data, it was determined that the victim was still alive at a given moment, because at that time, telephone contact had been established between the victim and a witness. (ECLI: NL: GHSHE: 2012: BX9271)

In another case, the court considered the alibi provided by a defendant's girlfriend implausible based on the historical data. The defendant claimed to be with his girlfriend at the time of the robbery that ended with fatalities. The court found this to be unlikely because historical data showed that the defendant had frequent telephone contact with his girlfriend on the night in question. (ECLI: NL: RBSGR: 2013: BZ0962)

In a hostage case it was assumed that a particular number was in use by the defendant based on historical traffic data. That particular number was used (frequently) to call the wife and family of a co-defendant. In addition, among other things, the transmission tower data showed that the number made contact with a tower in the vicinity of another co-defendant's parent's house. (ECLI: NL: RBUTR: 2012: BX5072)

In a raid on a house, detectives back track from a phone number that was used earlier, twice to call the victim. The number had apparently made contact with a tower in the vicinity of the victim's home, and was subsequently wire tapped, eventually leading to the suspect. (ECLI: NL: RBSGR: 2012: BX5776)

Unlike establishing the movements of suspects, or contacts between suspects, combining the time of a phone conversation with the duration thereof also can contribute to the evidence of a crime.

In one case of, amongst others, a threat, the victim claimed to recognize suspect's voice. Historical data showed that a short-term connection between numbers belonging to defendant and the victim had occurred. The short duration of the call (six seconds), corresponded to the expressed threat ('pay before July 1, or your head will be off!'). (ECLI: NL: RBAMS: 2012: BW3724)

The examples discussed above are all related to court rulings where historical traffic data were used as evidence, ending with a conviction.
Below we discuss some court rulings in which historical traffic data could not contribute to the evidence, in some cases even acquittal followed.

6.2.5 *Acquittals*

In a child abduction case, the court ruled that the reported account of events regarding the retrieval of children, times thereof, and a telephone call from the defendant from Belgium relating to that, were not plausible. The court found that the accused was elsewhere, which was supported by the historical data. (ECLI: NL: RBROE: 2012: BY0623)

A defendant is acquitted of a shooting. The court ruled there to be insufficient evidence of the defendant's involvement. The court took into consideration, amongst others, the fact that the accused might have, around the time of the shooting, had telephone contact with a possible co-defendant, which had led to the suspect's whereabouts at that time being noted in the case file. (ECLI: NL: RBSHE: 2012: BW2684)

Robbery of a jeweller: Although the court found that there were grounds for suspicion of suspect's involvement in two robberies of a jeweller, it ruled there to be insufficient evidence to prove such. The prosecutor deemed the defendant guilty, partly on the basis of available telecom data, which showed that around June 10, 2011, the defendant had had frequent contact with two accomplices who in the meantime had been convicted of those crimes. The prosecutor held that for both robberies the same method was used, namely three robbers, the same jeweller and the same escape route. During the trial, the defendant denied a particular mobile phone number to be his. The court held that there was insufficient evidence that this number indeed was in suspect's use. Moreover, it was shown that though the number did make contact with transmission towers near Dongen, contact had in fact not been made with a tower in Dongen itself. (ECLI: NL: RBBRE: 2012: BX8759)

The court acquitted a defendant of complicity in arson on a mosque. The fact that the defendant in question was implicated in the offence rested mainly on the statement of the co-defendant. The court held that the camera images stored in the phone and the historical data in the file, did not contradict the defendant and co-defendant's story. (ECLI: NL: RBSGR: 2012: BX7529)

A defendant was acquitted of four robberies. The historical traffic data suggested that one of the defendant's phones made contact with a tower a few blocks away from the crime scene, about ten minutes after the time of the robbery. The Court noted that the defendant could only be situated near the site of the robberies, as well as the fact that the defendant lived in Apeldoorn. For that reason it was not deemed illogical that the defendant's phone made contact with a tower in Apeldoorn, which, according to the court, yielded insufficient evidence to link the defendant to the robberies. (ECLI: NL: RBZUT: 2012: BW9618)

Finally, we mention a case in which the court explicitly doubts about the reliability of the historical traffic data:

In a triple murder/manslaughter a suspect is sentenced to life imprisonment. The results of the investigation into the mobile phone of the defendant could not be used as evidence (but weren't necessary), because the court held these data to be false. (ECLI: NL: RBDOR: 2012: BX9919)

6.3 Internet traffic data

When searching court rulings using the term IP address, a number of cases came up of which was insufficiently clear whether the IP address was actually supplied by a provider or by a site such as Marktplaats.nl (the Dutch equivalent to eBay). These rulings were thus disregarded from our study.
Of the 26 court rulings we found, dating between January 2009 and February 2013, the number of child pornography cases was quite noteworthy. More than fifty percent (15) of the court rulings pertained to downloading/distributing child pornography. Other types of offenses only occurred incidentally,

ranging from stalking and threats to fraud. The functional evidentiary value of internet traffic data appears to be different to the value of historical data shown above. The court rulings involving the use of internet data are distinguishable from those mentioned earlier in that the whereabouts of the suspect were not the main value, but rather because the means used more were more instrumental (content wise) in committing the offense, therefore providing substantive information. Such can be seen in the following examples. Somewhat analogous to the previous example, in which a relationship was established between the duration of a call, and the expressed threat in that conversation, with similar expected duration.

6.3.1 *Child pornography*

> *A defendant is convicted of possession of child and animal pornography. Based on information provided by Interpol, an investigation was conducted to determine which files had been downloaded to IP addresses. The so-called log files revealed that the IP address being used belonged to the defendant. (ECLI: NL: RBUTR: 2012: BW8244)*

Several court rulings leading to conviction, revealed a similar method of identifying a suspect on the basis of information provided by Interpol. Generally this information came from different countries. In one serious case of child pornography, the information came from Brazil, where a suspect spread photographs as a user of a peer-to-peer program. In one case, legal grounds led to an acquittal, which we will discuss below.

> *An investigation of an IP address was initiated after Interpol provided information about child pornography. The IP address appeared to be registered in the name of person X. In addition, seven computers appeared to be linked to this (office) network. The suspect admits to have used one of those computers to search for child pornography, on the basis of which the court assumes that the defendant downloaded the images he had looked at. Due to the fact that at the time the defendant was charged, viewing images (without storing them) was not illegal, the defendant is acquitted. (ECLI: NL: RBSGR: 2012: BV2841)*

6.3.2 *Advertisements*

There are different examples of how internet advertisements play a role in a criminal offence or in various kinds of criminal offences. Firstly, there is a case of large-scale fraud where goods were offered on sale via Marktplaats.nl (the Dutch eBay equivalent). The payments were received through accounts of straw men. The contracts between the suspect's IP addresses related to the ads, contributed to the body of evidence for the (many) offences. (ECLI: NL: RBSGR: 2012: BV2841)

Another ad related to the sale of sleeping pills.

A suspect and a co-offender were convicted of selling sleeping pills and sedatives in a business operation without a licence. The evidence was based in part on the internet advertisements, for which a particular IP address had been used. As a result of a search of the home on March 9, 2011, a laptop was found, which later turned out to have the same IP address. (ECLI: NL: RBSGR: 2012: BX4547)

A defendant is convicted of human trafficking with an accomplice. The defendant had also instigated a minor to prostitution. To this end, advertisements were placed on the internet, which could be traced back to the defendant's IP address. His defence that someone else had used his computer was found by the court to be unlikely for several reasons. (ECLI: NL: RBSGR: 2012: BW5833)

An advertisement/invitation for sexual encounters ended badly for those who responded. After the sexual encounters, the author of the advertisement, a participant in the sexual acts, accused the participants of rape. On the basis of information supplied by the provider, the ads could be traced to the defendant's IP address. She was convicted of making false accusations. (ECLI: NL: RBZUT: 2011: BR3110)

6.3.3 Threats

Threats made via the internet can also be traced to a perpetrator.

In one case of harassment/defamation, a mayor received messages sent from various public computers, including from a library. The court convicted the defendant for sending the messages that could be attributed to him, on the basis of a similar method on the internet, and the fact that the defendant frequented those places regularly to surf the internet. (ECLI: NL: RBUTR: 2012: BV7040)

In a case where a threat posted on the internet, in conjunction with a shooting at a school, investigators were led to the unprotected internet connection of the defendant's neighbours. (ECLI: NL: RBBRE: 2010: BO3363)

A defendant was sentenced to four years imprisonment for tax fraud and forgery in an organized context. The fact that the VAT declarations were sent from an IP address registered to the defendant's company, served as (part of the) evidence. (ECLI: NL: RBSGR: 2012: BX0774)

Finally, we will provide a few examples of other offenses whereby suspects could be tracked down by traces they left on the internet.

A teacher was convicted of sexual abuse of an underage student. The fact that computer messages were sent from the defendant's IP address was deemed relevant by the court. The defence that someone else had sent the messages is ruled by the court to be implausible. (ECLI: NL: RBSHE: 2012: BV8201)

A defendant was convicted for large-scale human trafficking from Iran. The defendant escorted people from Iran to the United Kingdom. The tickets were ordered via the internet. The payment by VISA, an email address used and an IP address were linked to an address in the Netherlands, which the court ruled to be directly linked to the defendant. (ECLI: NL: RBMAA: 2011: BQ8509)

A defendant was convicted, amongst other things, of preparatory acts of robbery with violence. Via an IP address traced to the suspect's home, it was established that the defendant had engaged in a MSN chat concerning the delivery of a firearm. Furthermore, further investigation of the internet address reveals information about burglaries, robberies, and ram raids. (ECLI: NL: RBUTR: 2012: BX2092)

6.4 In sum

It appears that traffic data can play a substantial role in the body of evidence for criminal offences. This has been demonstrated by the court rulings shown. We use the term substantial with reason. Court rulings include how traffic data, and the inferences that can be made from them, have contributed to sentencing. In some cases, the judge indicated that the traffic data yielded insufficient evidence. This was generally only the case when the traffic data was the only data to position a suspect at the crime scene. It has been argued, both in the literature as well as in political debate, that a suspect's innocence could also be proven using traffic data. Of this we found one example in our study of court rulings. Of course it is possible that such instances occurred at an earlier stage, for example during the investigation. Although the value of using traffic data can be clear from the court rulings discussed in this study, they shed no light on the age of the requested data. What is clear, is that the value of traffic data doesn't necessarily decrease as the data request covers a period further back in time.

Traffic data can contribute to evidence in different ways. Firstly, suspects can be positioned at a particular location or be linked to possible accomplices. This is the most common. In addition, incriminating connections with victims and other third parties can be established. When using the internet to commit crimes, the use of traffic data is often directly related to the offense. In other words, where the traffic data about telephone contacts between suspects shows that suspects knew each other, in itself this says nothing about the crime that was committed. On the other hand, the downloader of child pornography or those committing fraud via Marktplaats.nl, directly links the use of internet to the crime, assuming identification of the suspect is possible assuming a single user of the computer. In those cases, the computer or the internet are instrumental in the execution of the crime and much further evidence is often no longer necessary.
This may indicate that internet traffic data are primarily requested and analyzed for internet crimes and less used as supporting evidence to establish daily suspects' contacts or locations in other types of crimes.

7 Concluding remarks

In this report, we examined the Data Retention Act and the way in which telephone and internet traffic data are used in the criminal investigation practice. The study provides a view into the implementation of the retention act and shows how the retention act functions three years after its implementation. To do this we examined how traffic and location data are used in practice, for the detection and prosecution and as evidence of crimes. We also examined how often traffic and location data are requested and how monitoring compliance with the law is regulated.
The findings are based on a literature review as well as on interviews conducted with investigators, public prosecutors, lawyers, telecom service providers, regulators and other persons professionally dealing with the retention of telecommunications data.

Telephone Traffic Data
Historical traffic and location data on telecommunications are widely requested and analyzed for the purpose of criminal investigations. Professionals and experts value the possibilities of these data for investigative purposes. Telephony traffic and location data are requested particularly for the mapping of networks and for locating phones. The interviewed professionals regard the retrieval of traffic data as an investigative means that can be used in various ways in different kinds of cases. However, many professionals and experts believe that working with traffic and location data is specialist work. In practice, the data can result in complex questions, and the variety of new gadgets and technological developments do not make things any easier. Traffic data cannot only be seen as an important investigative tool, but can also play a role in the body of evidence. An analysis of published court rulings between July 2012 and February 2013 shows that judges have used telecommunications data as evidence in different situations, and have used it to motivate their judgements.

Internet traffic and location data
An analysis of telephone and internet traffic and location data provides a detailed picture of a person's life. On the other hand, it's an illusion to think that the analysis of traffic and location data can provide a complete picture of a person's communication behaviour. Nowadays, more and more communication takes place via the Internet, in many cases, with the use of services that don't fall under the jurisdiction of the Dutch Data Retention Act. The retention act is confined to within the Dutch boarders, whereas the Internet is not limited to one location or restricted by borders. This causes friction. Professionals and experts regularly involved in criminal investigations involving suspects who communicate via Internet, indicated that they would like to have access to the traffic data of these online communications. However, the number of Dutch data requests from foreign companies providing popular

internet services, lags far behind compared to neighbouring countries. Why that is, is unclear.
Our interviews revealed that the historical traffic and location data pertaining to internet, as defined in appendix B to section 13.2*a* of the Dutch Telecommunications Act, is not often used. The interviewed professionals and experts indicated that a significant portion of the retained and stored internet data is of limited value in investigations of newer forms of crime in which the internet plays a role. For this reason, data retained under the current directive are never or hardly ever requested.
At the same time, however, it appears that the degree of expertise in the police corps is insufficient to be able to fully exploit the available IP traffic data. It's complex material, and internet and the optimal use of digital data are not yet part of the daily practice of those interviewed, who engage in the investigation and prosecution of criminal offenses.
Apart from localizing a phone on the basis of the traffic data, ascribing names to IP addresses appear to be of great value to the investigative services. The police regularly want to know who the user of a particular IP address is. Identifying persons using a fixed internet IP address registered with a provider is relatively easy, because the retention act directs these data to be retained. But identifying users of mobile internet regularly poses a problem for which the retention act doesn't provide a solution. This is because many users log on to a so-called hotspot. This is a Wi-Fi network in which devices can log on to the Internet. With a Wi-Fi network, all users are grouped under one IP address, in order to use the Internet. Identifying individual users is thus not possible. Incidentally, many Wi-Fi network services are not covered by the current data retention act, because they cannot be classified as being public.
In addition, the shortage of IPv4 addresses appears to pose a growing problem. Because of this shortage, providers allocate IPv4 addresses to multiple customers simultaneously, causing difficulties for the identification of an individual user. As a result, the retention act is becoming increasingly inadequate in providing the data needed for the identification of an IP address user. This is not surprising, since the Telecommunications Act was written at a time when the world of telecommunications was very different to what it is now. The list of data to be retained, corresponding to Article 13.2a of the Telecommunications Act, is out-dated and has been overtaken by technological developments. This has resulted in the storage of sensitive civilian data, which is not or hardly ever used by investigative services. A reconsideration of the legislation on IP traffic data, and the retained data seems to be in order.

Legal safeguards/privacy
Mobile telephony has really taken off and nowadays many people have a smartphone. This multifunctional device, that also provides access to the

Internet, has become an essential part of everyday life. For this reason, the degree to which analysis of telephone and internet data can provide insight into a person's life has increased significantly in recent years. On the basis of traffic data, it is possible to outline the social network in which someone operates. The recorded location data can provide insight into where someone was at the time of communication using that device. When communicating with a smartphone, the retained data not only provides information about with whom, for how long and from which location contact was made, but also at what times and from which locations contact is made with the Internet. Hence, the privacy sensitivity of traffic and location data has increased over the years.

Storing this data can infringe on individual privacy in two ways. Firstly, the risk that unauthorized persons obtain access to these data, through for example hacking, increases merely by storing it. A second and different type of infringement occurs when law enforcement officials gain access to retained data in the context of criminal investigations. In the year 2012, the National Interception Unit processed 56,825 disclosure claims. However, due to their method of administration, there is no information about the number of times that data was actually delivered, the number of persons for whom a data request was made, or the number and type of cases in which this investigative instrument was used. It is therefore difficult to determine how many people's privacy has been infringed upon by using this means of investigation. It is also not clear how many of the claims made indeed pertained to data subject to retention under the so called act. In order to gain more insight into that, the National Interception Unit's managing systems require adjustment, such that the relevant data can be generated reliably. The Platform for Electronic Data Retention for the Investigation, Detection and Prosecution of Serious Crime's publication may be a helpful tool in redesigning the database and in publishing the annual figures.[116] There it is advised not only to include the number of claims, but also to count the number of responses from providers, by keeping track of how often requested information cannot be traced as well as responses that do not contain any information.

A frequently mentioned alternative to the retention of traffic data, where the privacy violation is less extensive, is freezing targeted data. However, this appears not to be a real viable alternative to the retention directive, because requesting older data is not possible. In order to use such data, one needs to know in advance that the data will be needed at a later date, so that the data can be frozen and thus remain available for use. Because crimes don't always come to the police's attention immediately, and suspects are often only identified (much) later, it is necessary to retain data so that they can later be used for investigative purposes.

116 http://ec.europa.eu/dgs/home-affairs/what-we-do/policies/police-cooperation/data-retention/experts-group/docs/guidance_on_statistics_position_paper_16_datret_final_en.pdf (consulted on March 20, 2013).

To ensure that the data is stored, protected and disposed of properly, according to the Data Retention Act, and thus that the risk of unauthorized access to the data is reduced, providers and telecom companies have been legally required to take technical and organizational measures in order to prevent abuse of the stored data, and they have been required to destroy the retained data after the retention period. In addition, customers of Internet Service Providers have the right to access the data retained about them. However, the two requests for access to private traffic and location data submitted for this study were barely honoured.
Monitoring compliance with the obligations for Internet Service Providers and telecommunications companies is at first the responsibility of the Telecom Agency, and ultimately of the Minister of Economic Affairs. The Data Protection Agency also plays a role in monitoring and oversees all legislation involving the retention, use, or processing of personal data.
Access to and the use of traffic data by criminal investigators is indisputably an infringement of the privacy of citizens. This infringement must be deemed necessary and meet requirements of proportionality and subsidiarity. The Code of Criminal Procedure thus regulates what conditions need to be met and who can access retained telecom and internet data. Moreover, it stipulates that persons whose traffic data has been requested, must be informed as soon as possible, without compromising the investigation. However, current draft legislation[117] proposes to abolish the notification requirement for requests of traffic data, because it is assumed that this authority poses 'a relatively light infringement' on the privacy of individuals. This reasoning is at odds with the fact that in recent years, due to the increased use of the mobile phone, an analysis of telephone and internet traffic data can in fact provide an increasingly detailed insight into someone's private life.

Storage, security and monitoring
The retention directive pertains to the storage of privacy sensitive data from which information can be derived regarding the residence and contacts of individuals. This makes proper monitoring of the implementation of the Data Retention Act a necessary given, in order to prevent data being misused or misappropriated.
The storage of traffic data directed for retention by the Data Retention Act, is decentralised in the Netherlands, so Internet Service Providers themselves are responsible for the local storage, the data security and the ultimate destruction of the data. The relevant data can be requested by investigative services for law enforcement purposes, from the providers through the National Interception Unit. The telephone data can, in the Netherlands,

117 Draft legislation for the amendment of the Dutch Code of Criminal Procedure and the Code of Civil Procedure in light of increasing the performance ability of the police. See www.rijksoverheid.nl/documenten-en-publicaties/kamerstukken/2012/09/03/consultatieservice-conceptwetsvoorstel.html (consulted on May 1, 2013).

be requested up to a year prior to the date of the application. Internet data can be made available up to six months prior to the request date.
The Telecom Agency acts as supervisor, but only has been endowed the authority to monitor the correct implementation of business processes. The Telecom Agency has no authoritative powers to monitor the content of the retained data. When a government decides to store sensitive private information of citizens, it also has the responsibility to make sure that effective supervision is realised for both the content of the data stored as well as the data that are ultimately supplied to the investigative services. Currently, however, the Telecom Agency is unable to view and monitor the content of traffic data supplied under Article 18.7, section 2 of the Telecommunications Act. It is thus recommended that the role of the regulator is expanded accordingly.

The data retention periods
The retention periods for telephony traffic and location data is one year. This period was found to be very workable by the interviewed professionals and experts. In some cases, a year is too short to carry out an investigation well, but this proved to be the exception rather than the rule, and the interviewed professionals and experts wonder how long data should be retained in order to resolve these issues. Ultimately, it pertains to sensitive information of which it is undesirable to store anywhere for an unnecessarily long period of time.
The retention period for historical internet traffic data is six months. It is not quite clear on the basis of which arguments the retention periods for telephone and internet traffic data vary. It is possible that arguments pertaining to privacy issues (in part) underlie this distinction. However, the nature of the internet data stored, effectively doesn't pose a greater infringement on individual privacy when compared to the nature of telephony data. Regarding internet data, only information indicating the location of an internet contact can be derived, and which IP address was used is retained. Only when retrieving email traffic, it is possible to see with whom contact has been made. This is in contrast to data on telephone traffic, for which, when a phone is used, a social network can be mapped.
The retention period of six months for internet data is considered unanimously to be too short by the criminal investigations professionals and experts. This is particularly so for complex cases where such data can be useful. According to the interviewed professionals and experts the data is not available long enough. This mainly concerns data that can be used to identify the user of an IP address.
Given that the volume of internet traffic will increase while telephone traffic is likely to decrease further in the future; and given that under the current regulations there doesn't appear to be a clear difference in the degree to which stored telephone and internet data infringe on the privacy of individual citizens; and given that their use in criminal investigation practices is also

similar (especially in the case of smartphones), it would seem obvious to harmonize these retention periods. The harmonization of the retention periods would provide a solution to a number of technical and practical problems that providers currently face. Moreover, the law could then be applied similarly for all types of mobile or prepaid internet. Current legislation leads to inconsistent policies when it comes to internet traffic data generated by prepaid cards versus data generated by subscribers. A retention period of six months currently applies to subscribers whereas for prepaid card users a retention period of one year is maintained. This difference can be ascribed to the fact that legislation on the retention periods for prepaid telecommunications data is regulated in Article 13.4 section 3 of the Telecommunications Act.

For various telephony services, such as traditional telephony, mobile telephony and internet telephony (VoIP), which the government considers similar in terms of functionality, a retention period of twelve months is maintained. This is so despite the fact that strictly speaking, internet telephony yields internet traffic data, for which six months retention is mandated. This is confusing, particularly because other voice services available via the internet do hold the shorter retention period of six months. As such, providers of communication services are not always sure whether they are obliged to store data.

General

Using traffic data in criminal investigations can be characterized as a 'traditional' way of investigating. The current investigation practice still relies heavily on this traditional investigative method, which incidentally still yields a lot of useful contact information.

Due to ongoing technical developments, the current retention directive for internet traffic data can by and large be considered to be obsolete. The law was drafted in a period when one logged onto the internet with a modem, while nowadays, many people are online 24 hours a day, 7 days a week. The list of data to be retained mentioned in the appendix to Article 13.2*a* of the Telecommunications Act, is dated. As a result, the private data that is being stored, is hardly being used by investigative services. This is an undesirable situation.

A strict distinction between telephony and internet as maintained under the current Retention Act is outdated and results in ambiguities and technical and practical problems. A careful review of the regulations, retention periods and in particular which IP traffic data should be preserved seems to be necessary. Inextricably coupled with an adjustment of the to be retained internet traffic data, is the need for a careful reconsideration of the retention period. The privacy sensitivity of the information to be included in the new data retention laws should nevertheless play a decisive role.

Having noted that, however, it seems that retrieving telecommunications data will become less useful in the future. Other forms of communication, such as communication through social media and through games, are not covered by the Retention Act and cannot be traced. The current legislation does not reflect these developments and it is unlikely that 'local' Dutch legislation can overcome this and can be meaningful in the borderless and location free space of the internet.

Given the international nature of many forms of crime, European harmonization on retention periods and the retrieval of data is desirable. European harmonization does not, however, resolve all the possible challenges presented by the virtual world. Security minded thinking stimulates an expansion of retention directives to involve more possibilities for the use of internet data in the investigation of crime. However, from the perspective of the privacy of citizens, this is an undesirable development. Moreover, the substantive availability of personal information will not only be desirable for investigative authorities, other parties may also be eager to gain access to such data. The retention of such data thus carries a greater risk of abuse. In addition, having more information doesn't necessarily mean that the investigation process will be more efficient, as not all data is useful, thus potentially impeding efficient investigation. The search for an alternative to the current data retention is a hefty challenge, with on the one hand the interest of internet communication for crime investigation, and on the other hand the interest of minimal privacy infringement. Developing investigation possibilities to combat new forms of crime, will give us challenging and complex issues to tackle in the future, which time and again need to be considered carefully.

Literature

Boot, R., Bosch, J. van der, Vervaet, E., & Varkevisser, K. (2006). *Onderzoek naar de nationale Implementatie van de Europese richtlijn dataretentie. [Research on the national implementation of the European Data Retention Directive].* Verdonck, Klooster & Associates. Accessed on June 1, 2013: www.eerstekamer.nl/behandeling/20081209/onderzoek_naar_de_nationale/f=y.pdf.

Central Government (s.a.). *Telecom Data for Criminal Investigations.* Accessed on June 1, 2013: www.rijksoverheid.nl/onderwerpen/telecom gegevens-voor-opsporing.

Central Government (2012). *Draft Bill to Amend the Criminal Code and the Code of Civil Procedure Concerning the Performance Enhancement of the Police.* Accessed on May 1, 2013: www.rijksoverheid.nl/documenten-en-publicaties/kamerstukken/2012/09/03/consultatieversie-concept-wetsvoorstel.html.

DatRet/Expgrp (2009). *Position Paper No 5. Closer understanding of the term 'Internet Telephony' in relation to its application in Directive 2006/24/EC.* Accessed on March 20, 2013; http://ec.europa.eu/dgs/home-affairs/what-we-do/policies/pdf/series_a_position_paper_5_final_14_07_2010_closer_understanding_of_the_term_internet_telephony_en.pdf.

DatRet/Expgrp (2012). *Position Paper No. 16: Guidance on the Member States obligation to submit to the Commission annual statistics pursuant to Directive 2006/24/EC.* Accessed on March 20, 2013: http://ec.europa.eu/dgs/home-affairs/what-we-do/policies/police-cooperation/data-retention/experts-group/docs/guidance_on_statistics_position_paper_16_datret_final_en.pdf.

Electronic Frontier Foundation (2012). *Mandatory Data retention: Europe.* Accessed on December 7, 2012: www.eff.org/issues/mandatory-data-retention/eu.

European Commission (2011). *Verslag van de Commissie aan de Raad en het Europese Parlement. Evaluatie van de richtlijn gegevensbewaring (Richtlijn 2006/24/EG) [Report from the Commission to the Council and the European Parliament. Evaluation of the Data Retention Directive (Directive 2006/24/EC)].* http://eur-lex.europa.eu/LexUriServ/LexUriServ.do?uri=COM:2011:0225:FIN:nl:PDF.

Frost & Sullivan (2010). *Meeting the challenges of data retention: Now and in the future.* Accessed on December 1, 2012: www.frost.com.

Google. *Transparancy Report.* Accessed on March 13, 2013: www.google.com/transparencyreport/userdatarequests/legalprocess.

Google. *Transparancy Report.* Accessed on March 19, 2013: www.google.com/transparencyreport/userdatarequests/NL.

Hathaway, M.E. & Klimburg, A. (2012). *Preliminary considerations: On national cyber security.* In Klimburg, A. (ed.), *National Cyber security framework manual* (pp. 1-43). Tallin, Estland: NATO Cooperative Cyber Defence Centre of Excellence.

Hustinx, P. (2010). *The moment of truth for the Data Retention Directive.* Presentation at The 'Taking on the Data Retention Directive' Conference, Brussel, December 3, 2010.

ICT-recht (2011). *Overzicht bewaarplicht: wie wel en wie niet [Overview Retention: Who should and who shouldn't].* Accessed on April 4, 2013: https://ictrecht.nl/ictrecht/overzicht-bewaarplicht-wie-wel-en-wie-niet.

Independent Post and Telecommunications Authority (OPTA 2009). *Jaarverslag en Marktmonitor [Annual Report and Market Monitor].* The Hague: OPTA.

Independent Post and Telecommunications Authority (OPTA 2010). *Toezichtsactie registratieplicht bij 15 internetleveranciers in de hotelbranche [Monitoring Action registration requirement at 15 internet providers in the hotel industry].* Accessed on December 11, 2012: www.opta.nl/nl/actueel/alle-publicaties/publicatie/?id=3296.

Independent Post and Telecommunications Authority (OPTA 2011). *Presentatie Markt Monitor [Presentation Market Monitor 2010].* The Hague: OPTA.

Independent Post and Telecommunications Authority (OPTA 2012). *Marktcijfers tweede kwartaal [Market figures second quarter].* The Hague: OPTA.

ITU (International Telecommunications Union) (2011). *World telecommunication/ICT indicators database* (15e editie).

Knol, P.C., & Zwenne, G.J. (2013). *Tekst en commentaar 'Telecommunicatie- en privacyrecht' [Text and Commentary 'Telecommunication Privacy Law'].* The Hague: Wolters Kluwer.

Koops, B.J. (2002). *Strafvorderlijk onderzoek van (tele)communicatie 1838-2002: Het grensvlak tussen opsporing en privacy [Criminal investigation of (tele) communication 1838-2002 : The interface between detection and privacy].* Deventer: Kluwer.

Koops, B.J., Bekkers, R., Bongers, F., & Fijnvandraat, M. (2005). *Aftapbaarheid van telecommunicatie: Een evaluatie van hoofdstuk 13 Telecommunicatiewet [Legal interception of telecommunications : A Review of Chapter 13 Telecommunications].* Tilburg: TILT – Centrum voor Recht, Technologie en Samenleving (Centre for Law, Technology and Society).

Koops, B.J., Leenes, R., Hert, P. de, & Olislaegers, S. (2012). *Misdaad en opsporing in de wolken: Knelpunten en kansen van cloud computing voor de Nederlandse opsporing [Crime and detection in the clouds: Problems and opportunities of cloud computing for Dutch criminal investigations].* Tilburg: TILT – Centrum voor Recht, Technologie en Samenleving (Centre for Law, Technology and Society).

Mevis, P.A.M. et al. (2005). *Wie wat bewaart heeft wat: Onderzoek naar nut en noodzaak van een bewaarverplichting van historische verkeersgegevens van telecommunicatie [He who keeps something, has something: Research into the usefulness and necessity of a retention directive for historical telecommunications traffic data].* Rotterdam: Erasmus University.

Munnichs, G., Schuijff, M., & Bestters, M. (2010). *Databases: Over ICT-beloftes, informatiehonger en digitale autonomie [Data bases: On ICT-promises, hunger for information and digital autonomy]*. The Hague: Rathenau Institute. www.rathenau.nl/publicaties/publicatie/databases-over-ict-beloftes-informatiehonger-en-digitale-autonomie.html.

Nelen, H., Leeuw, F,. & Bogaerts, S. (2010) *Antiterrorismebeleid en evaluatieonderzoek: Framework, toepassingen en voorbeelden [Anti-terrorism Policy and evaluation: Framework, applications and examples]*. The Hague: Boom Legal Publishers.

Odinot, G., Jong, D. de, Leij, J.B.J. van der, Poot, C.J. de, & Straalen, E.K. (2012). *Het gebruik van de telefoon- en internettap in de opsporing [The use of Telephone and internet tapping in criminal investigations]*. The Hague: Boom Legal Publishers. Research and Policy 304.

Rathenau Institute (2013). *Handout ICT-committee Parliament.* Accessed on June 1, 2013: www.rathenau.nl/publicaties/publicatie/hand-out-ict-commissie-tweede-kamer.html.

Smith, B. (2013). *Microsoft releases 2012 law enforcement requests report.* Accessed on March 25, 2013: www.blogs.technet.com/b/microsoft_on_the_issues/archive/2013/03/21/microsoft-releases-2012-law-enforcement-requests-report.aspx.

Stratix Consulting (2009). *Grenzen aan aftapbaarheid? [Limitations to tapability?]* Hilversum: Stratix.

Telecom Agency (2010). *Eindrapport Nulmeting Wet bewaarplicht telecommunicatiegegevens [Baseline measure of the Telecommunications Data Retention Act]*. (final report) Accessed on June 1, 2013: www.eerstekamer.nl/behandeling/20100817/_eindrapport_nulmeting_wet/f=y.pdf.

Telecom Agency (2012). *Staat van de Ether [State of the Ether]*. Accessed on June 29, 2013: www.agentschaptelecom.nl/onderwerpen/frequentie-management/staat-van-de-ether.

Telecom Agency (z.j.). *Opslag Telecommunicatiegegevens [Storage of Telecommunications data]*. Accessed on April 4, 2013: www.agentschaptelecom.nl/onderwerpen/veiligheid/opslag-telecomgegevens.

TNO (Nederlandse Organisatie voor toegepast-natuurwetenschappelijk onderzoek, the Netherlands Organisation for Applied Scientific Research) (2010). *Marktrapportage Elektronische Communicatie [Electronic Communications Market Report]*. Delft: TNO.

TNO (Nederlandse Organisatie voor toegepast-natuurwetenschappelijk onderzoek, the Netherlands Organisation for Applied Scientific Research) (2011). *Marktrapportage Elektronische Communicatie [Electronic Communications Market Report]*. Delft: TNO.

Twitter (2012). *Information requests.* Accessed on March 25, 2013: https://transparency.twitter.com/information-requests-ttr2.

Wartna, B. (2005). *Evaluatie van daderprogramma's [Evaluation of perpetrator programs]*. The Hague: Boom Legal Publishers.
Zenger, R. (2011a). *Ik, in de ogen van T-mobile [Me in the eyes of T-mobile]*. Accessed on December 5, 2012: https://rejo.zenger.nl/focus/ik-de-ogen-van-t-mobile.
Zenger, R. (2011b). *Locatie te achterhalen uit call detail records? [Can locations be traced on the basis of call detail records?]*. Accessed on December 5, 2012: https://rejo.zenger.nl/focus/locatie-te-achterhalen-uit-call-detail-records.

Memoranda, minutes and decrees

Preparatory Memorandum (Senate), 2007/08, 31 145, B.
Preparatory Memorandum (Senate), 2008/09, 31 145, C.
Preparatory Memorandum (Senate), 2008/09, 31 145, F.
Preparatory Memorandum (Senate), 2008/09, 31 145, F (NMvA).
Preparatory Memorandum (Senate), 2010/11, 32 797, A.
Explanatory Memorandum (House of representatives), 1989/90, 21 551, nr. 3.
Explanatory Memorandum (House of representatives), 2006/07, 31 145, nr. 3.
Preparatory Memorandum (House of representatives), 2007/08, 31 145, nr. 9.
Preparatory Memorandum (House of representatives), 2007/08, 31 145, nr. 14.
Preparatory Memorandum (House of representatives), 2009/10, 32 185, nr. 2.
Explanatory Memorandum (House of representatives), 2009/10, 32 185, nr. 3.
Minutes (Senate), 2008/09, 39, p. 1808.
Minutes (Senate), 2008/09, 40, p. 1839 e.v.
Minutes (Senate), 2008/09, 40, p. 1845.

Staatsblad (2002). Decree of December 18, 2001, Concerning the rules for the collection of number data through the use of alternative frequencies and file analysis in order to investigate telecommunications (Decree special telecommunication number data collection). *Staatsblad (State Gazette)*, nr. 31.
Staatsblad (2004). Decree of August 3, 2004, concerning the guidelines for exacting user information and telecommunications traffic from a public telecommunications network provider and a telecommunications service (Decree exacting telecommunication data). *Staatsblad (State Gazette)*, nr. 394.
Staatsblad (2006). Decree of December 21, 2006, concerning the execution of directives relating to several provisions of the Dutch Code of Criminal Procedure regarding the investigation of terrorism. (Decree investigation of terrorist crimes). *Staatsblad (State Gazette)*, nr. 730.
Staatsblad (2011). Act of July 6, 2011 leading to the amendment of the Telecommunications Act concerning the adjustment of telecommunications data

retention periods, specifically pertaining to internet access, internet email and internet telephony. *Staatsblad (State Gazette)*, nr. 350.
Staatsblad (2012). Act of November, 30, 2012, leading to the amendment of the Police Act 2012. *Staatsblad (State Gazette)*, nr. 615.

Jurisprudence

European Court of Human Rights (ECHR)

ECHR December 4, 2008, nr. 30562/04 (S and Marper/The United Kingdom).

Court of Justice of the European Union (CJEU)

CJEU February 10, 2009, nr. C-301/06 (Ierland/Europees Parlement en Raad van de Europese Unie).

Court of Appeal

Court of Appeal The Hague, October 26, 2012, *ECLI:NL:GHSGR:2012:BY1648*
Court of Appeal 's-Hertogenbosch, May 30, 2012, *ECLI:NL:GHSHE:2012:BW7042*
Court of Appeal 's-Hertogenbosch, October 2, 2012, *ECLI:NL:GHSHE:2012:BX9271*
Court of Appeal Amsterdam, October 31, 2012, *ECLI:NL:GHAMS:2012:BY1810*
Court of Appeal Amsterdam, November 6, 2012, *ECLI:NL:GHAMS:2012:BY2562*
Court of Appeal Arnhem, June 18, 2012, *ECLI:NL:GHARN:2012:BW8652*
Court of Appeal Arnhem, August 30, 2012, *ECLI:NL:GHARN:2012:BX6113*
Court of Appeal Arnhem, August 30, 2012, *ECLI:NL:GHARN:2012:BX6121*

District Court

District Court The Hague, January 5, 2012, *ECLI:NL:RBSGR:2012:BW5833*
District Court The Hague, February 3, 2012, *ECLI:NL:RBSGR:2012:BV2841*
District Court The Hague, April 16, 2012, *ECLI:NL:RBSGR:2012:BX0774*
District Court The Hague, August 14, 2012, *ECLI:NL:RBSGR:2012:BX4547*
District Court The Hague, August 21, 2012, *ECLI:NL:RBSGR:2012:BX5105*
District Court The Hague, August 27, 2012, *ECLI:NL:RBSGR:2012:BX5776*
District Court The Hague, August 30, 2012, *ECLI:NL:RBSGR:2012:BX6147*
District Court The Hague, September 17, 2012, *ECLI:NL:RBSGR:2012:BX7529*
District Court The Hague, October 15, 2012, *ECLI:NL:RBSGR:2012:BY0109*
District Court The Hague, February 7, 2013, *ECLI:NL:RBSGR:2013:BZ0962*

£31.00

District Court 's-Hertogenbosch, March 8, 2012, *ECLI:NL:RBSHE:2012:BV8201*
District Court 's-Hertogenbosch, April 19, 2012, *ECLI:NL:RBSHE:2012:BW2684*
District Court 's-Hertogenbosch, October 19, 2012, *ECLI:NL:RBSHE:2012:BY0575*
District Court Alkmaar, August 14, 2012, *ECLI:NL:RBALK:2012:BX4768*
District Court Amsterdam, April 19, 2012, *ECLI:NL:RBAMS:2012:BW3724*
District Court Amsterdam, July 18, 2012, *ECLI:NL:RBAMS:2012:BX1952*
District Court Amsterdam, July 20, 2012, *ECLI:NL:RBAMS:2012:BX3164*
District Court Amsterdam, July 23, 2012, *ECLI:NL:RBAMS:2012:BX5674*
District Court Arnhem, November 14, 2012, *ECLI:NL:RBARN:2012:BY2895*
District Court Breda, November 9, 2010, *ECLI:NL:RBBRE:2010:BO3363*
District Court Breda, March 25, 2011, *LJN BP9283*
District Court Breda, August 9, 2012, *ECLI:NL:RBBRE:2012:BX4244*
District Court Breda, October 1, 2012, *LJN BX8759*
District Court Dordrecht, October 11, 2012, *LJN BX9919*
District Court Haarlem, April 19, 2012, *ECLI:NL:RBHAA:2012:BW2968*
District Court Maastricht, June 20, 2011, *ECLI:NL:RBMAA:2011:BQ8509*
District Court Noord-Nederland, January 24, 2013, *ECLI:NL:RBNNE:2013:BY9376*
District Court Roermond, October 12, 2012, *ECLI:NL:RBROE:2012:BY0623*
District Court Rotterdam, March 27, 2009, *ECLI:NL:RBROT:2009:BH9324*
District Court Rotterdam, July 12, 2012, *ECLI:NL:RBROT:2012:BX1291*
District Court Utrecht, February 27, 2012, *ECLI:NL:RBUTR:2012:BV7040*
District Court Utrecht, June 4, 2012, *ECLI:NL:RBUTR:2012:BW8244*
District Court Utrecht, June 29, 2012, *ECLI:NL:RBUTR:2012:BX2092*
District Court Utrecht, August 20, 2012, *ECLI:NL:RBUTR:2012:BX5072*
District Court Utrecht, October 9, 2012, *ECLI:NL:RBUTR:2012:BX9634*
District Court Zutphen, July 26, 2011, *ECLI:NL:RBZUT:2011:BR3110*
District Court Zutphen, June 27, 2012, *ECLI:NL:RBZUT:2012:BW9618*

Appendix 1 Advisory Board

Chairman

Prof. Mr. M.J. Borgers	Professor of Criminal (Procedural) Law, Free University of Amsterdam

Members

Prof. Dr. E.J. Koops	Professor of Regulating Technology, Tilburg Institute for Law, Technology and Society (TILT), Tilburg University
Mr. R. van Bosbeek	National Interception Unit, National Unit, National Police
Mr. L.J.A. van Zwieten	National Prosecutor Cybercrime, National Public Prosecution Service Rotterdam
Mr. A. Sternenberg	Main Security Vodafone, representation of Platform 13 (consultation between government and providers of telecommunications)
Mr. B.A. Stap	Ministry of Security and Justice Research, Office Platform Interception Decryption & Signal Analysis
Drs. C.J.A.M. Meijer	Ministry of Security and Justice, Fraud and Planning Section

£ 31.00